Quality Involvement Flow

The Systemic Organization

Quality Involvement Flow

The Systemic Organization

Domenico Lepore
Angela Montgomery
Giovanni Siepe

CRC Press
Taylor & Francis Group
Boca Raton London New York

CRC Press is an imprint of the
Taylor & Francis Group, an **informa** business
A PRODUCTIVITY PRESS BOOK

CRC Press
Taylor & Francis Group
6000 Broken Sound Parkway NW, Suite 300
Boca Raton, FL 33487-2742

Printed on acid-free paper
Version Date: 20160322

International Standard Book Number-13: 978-1-4987-5588-7 (Paperback)

Library of Congress Cataloging-in-Publication Data

Names: Lepore, Domenico, 1959- author. | Montgomery, Angela, author. | Siepe, Giovanni, author.
Title: Quality, involvement, flow : the systemic organization / Domenico Lepore, Angela Montgomery, and Giovanni Siepe.
Description: Boca Raton, FL : CRC Press, 2016.
Identifiers: LCCN 2016005362 | ISBN 9781498755887 (pbk.)
Subjects: LCSH: Management. | System analysis. | Organizational behavior. | Organizational change. | Quality control.
Classification: LCC HD31 .L3834 2016 | DDC 658--dc23
LC record available at http://lccn.loc.gov/2016005362

Visit the Taylor & Francis Web site at
http://www.taylorandfrancis.com

and the CRC Press Web site at
http://www.crcpress.com

This could have been a world where the nature of each thing would be deduced from its parts. A predictable, orderly world. A world devoid of wonder. And then we would say, "Things are this way because they must be this way."

Instead, at each step a whole new world emerges, one we could never have predicted from anything we knew before. Until we must conclude that our finite work somehow contains infinite possibilities, that both nothing and everything are possible.*

This book is dedicated to all those who refuse to accept things as they are, knowing we have a precise task: to unveil the infinite possibilities with courage and make sure that potential emerges into reality.

* Adapted from *Daily Dose* from Chabad.org edited by Tzvi Freeman, based on Sefer Ha-Ma'amarim 5715, pp. 38–47, a talk given by Rav Menachem Mendel Schneerson, the Lubavitcher Rebbe.

Contents

Preface

We wrote this book to present the elements of science and method that should inform the leadership of organizations in the twenty-first century. Our effort has been to outline a cohesive and comprehensive approach to this systemic understanding and to use these elements to provide a unified vision for succeeding in our age of complexity.

The thought process and the professional practice that inspired these pages have been profoundly influenced by the monumental work of W. Edwards Deming and Eliyahu M. Goldratt and they are informed by their scientific and epistemological paradigm.

The first seed for this approach was developed by Domenico Lepore and Oded Cohen almost 20 years ago and it was named *The Decalogue*, first published as *Deming-and-Goldratt: The Decalogue* (North River Press, 1999). Since then, further work on The Decalogue methodology has been carried out internationally in a wide range of organizations to refine its usefulness and applicability; in particular, the authors of *Quality, Involvement, Flow* derived great inspiration and insights from Network Theory and the light it casts on our understanding of the nonlinear dynamics that takes place in the highly interconnected world we now live in.

Such an unprecedented level of interconnectedness, the fast growing socioeconomic environment shaped by networks and the role that the web plays in our lives call for a major overhaul of the content and the pattern of our learning. Particularly, academic institutions responsible for educating the business leaders of tomorrow must awaken to the sheer inadequacy of existing programs and take action accordingly.

Quality, Involvement, Flow is not about business improvement, reengineering or any of the techniques aimed at generating some economic result. It draws heavily on the teachings of Dr. Deming and Dr. Goldratt and, as such, its goal is the transformation of the present style of leadership (and management) into one of whole system optimization. The transformation that we advocate cannot be undertaken without understanding and embracing a basic set of values, and it is mandatory to realize that the achievement of economic results must be connected with a precise ecology of the mind. In this way, the legitimate pursuit of personal monetary wealth will not be disjointed from the quest for purposefulness that should be the prime motivator for any human endeavor. This is the transition that leads to long-term, sustainable results.

This book aims at being both a cognitive challenge and an operational guide; it should be a positive stimulus for acting in accordance with the emerging, network-shaped reality of our world. It contains little redundancy and requires from the reader a nonnegligible amount of focus and intellectual openness. It will not provide suggestions, tips, and quick fixes and it will, at times, feel uncomfortable.

In a world that is drowning in a chaotic swirl of shallow information, instant pudding, and evocative feel-good calls to "systems thinking," *Quality, Involvement, Flow* strives to offer an orderly and rigorous knowledge-based reference point for the new leaders of the millennium.

Acknowledgments

This book would never have existed without the fundamental work on management developed by Dr. W. Edwards Deming and Dr. Eliyahu Goldratt. They have informed and shaped our views on economics and management and we are forever indebted to them.

We are ever grateful to our friend and mentor Oded Cohen for his painstaking and selfless teaching. He is one of the foremost experts in the world in the Theory of Constraints and we have enjoyed the great good fortune of his input, guidance, and friendship over the years.

Whatever we have written, or ever will write on the subject of variation owes a debt of gratitude to Dr. Don Wheeler. We thank him for his unique mastery in understanding variation and the unsurpassed clarity of his writing.

We thank Bob Sproull for his spontaneous generosity in connecting us with Michael Sinocchi, our editor. Bob is an example of how positive collaboration can create new possibilities in our interconnected world.

It is thanks to Michael Sinocchi of Productivity Press that we have had this opportunity to give ample expression to our systemic approach to management developed in over 20 years of work on the field. We are grateful for his enthusiasm, encouragement, and openness to a body of work that is rigorous but also unconventional and, at times, challenging.

Dr. Anthony Masys at Springer also receives our thanks for believing in our work and our crusade to bring systemic management to a world plagued by silos.

Our work has been made all the more enjoyable thanks to our friendship with Professor David Strong in Victoria, BC, a true scientist with the ability to think beyond boundaries. Deep gratitude goes to Adam Sharpe CPA, also in Victoria, who ennobles his profession, is a true mensch and has been like a brother to us.

In Italy, we thank Livio Brachetti for having the vision, patience, foresight, and talent to create a platform that can truly improve the nature of relationships on the web, and also for embracing the way of systemic management and for inviting us to be his comrades in Social Nation. Our thanks to Stefano Righetti of Hyphen-Italia for relentlessly and courageously pursuing intelligence, ethics, and beauty in his entrepreneurial achievements in a world often dominated by a lack of thinking and platitudes. Stefano understands the Decalogue as applied to the supply chain in a way that few others do. We are grateful to both Livio and Stefano for their warmth and friendship.

In the last decade, we have made an increasing effort to understand the cultural roots, context, and more profound implications and application of the Theory of Constraints. In this endeavor, in ways that we are not sure we can adequately express, we have been profoundly influenced by the body of work of Rav Menachem Mendel Schneerson, the Lubavitcher Rebbe. The entire philosophical framework and value system that has inspired this book is based on our understanding of his insights. The world continues to benefit from his scholarship, wisdom, profound understanding of human nature and unprecedented intuition for network organization. Above all, we are constantly inspired by his relentless call for “ufarazta,” or breaking through barriers to achieve what only seems impossible when we fail to see the interdependencies at every level, physical and otherwise. Among the people who helped us gain some understanding of the Rebbe’s monumental contribution to the betterment of mankind, we would like to mention Rav Chaim Miller, for his exceptional scholarship and ability to innovate within the boundaries of tradition; Rav Motti and Shterni Seligson who through their work and actions serve as role models for a new generation; and Rav Meir Kaplan for the depth and punctiliousness of his teaching. Our lives are enriched by their knowledge and friendship.

Chapter 1

Navigating the New Complexity

The greatest challenge for organizations today is complexity. It is here to stay, and leaders and managers need to learn to navigate and lead in a new and complex reality. A major difficulty is that there is little real understanding in the corporate world of what complexity is, how it originates, and what deep shifts, not just techniques and tactics, are required to operate and innovate in a complex world.

When Christopher Columbus and other explorers discovered new worlds, it was thanks to their ability to navigate oceans. This ability was based on knowledge of astronomy and trade winds, and mariners could only rely on rudimentary instruments like the compass, the sextant and nautical charts. In the twentieth century, travelling to the moon and back became possible, thanks to the new knowledge supplied by the theory of relativity and the technology available to build a space shuttle.

In the twenty-first century, whatever an organization does, from mining to healthcare, it has to be able to steer through complexity and many leaders are lagging behind. This is because the way organizations are structured and the way they operate is still based on models that are now out of date. Nobody would dream of travelling into space without the right vehicle and equipment and without astronauts who know exactly what to do. Why, then, do so many still try to manage organizations the way they were managed a 100 or even 50 years ago? Why have the programs in business schools evolved so little in recent decades?

The problem is that organizations and business schools are still very much trapped in an outdated paradigm of silos, fragmentation, conflicts, and a zero sum game. There has not yet been a real shift away from an outdated mechanistic worldview of organizations toward an urgently needed systemic understanding. No matter which techniques forward-looking managers may try to adopt, management remains widely rooted in the idea of command and control, and this is reflected in a traditional, hierarchical/functional organizational design (silo mentality).

The reason why this is a problem is that it imposes from the outside a structure that does not reflect the inherent nature of the organization as one of *interactions*. Lip service to more enlightened ways is not enough. In order to navigate complexity, we need to leave behind the mechanistic approach that was suitable in the nineteenth and early twentieth century and truly understand what complexity is. If not, we end up in the paradoxical situation of trying to manage complexity by "simplifying" it. We end up using even the wrong language, the way some world famous top advisory firms do, when they talk about "confronting" or "cutting through" complexity and "breaking it down into parts." We will soon see why this is completely inappropriate.

To find a way to manage complexity, we need to strip back the organization to its essential nature so we can organize and manage it appropriately. We need to equip ourselves with adequate knowledge, methods, and tools. To borrow a term from a famous movie about the space program, if we are to navigate our organizations through complexity, we need "the right stuff," at the theoretical, methodological, and human level. The purpose of this book is to identify that right stuff.

1.1 Dividing Things Up Is No Longer the Answer

Most people desire a way to simplify what they do. Who can blame them? We are constantly bombarded by a myriad of inputs. E-mail boxes overflow and there is no end to the number of "pieces" in the puzzle that every leader and manager faces on a daily basis. Technology imposes continuous updates. The defensive approach that many choose is to "divide and conquer." In other words, faced with a multiplicity of inputs, the reaction is to try and control them by breaking them up into smaller "pieces." The assumption here is that by breaking a big situation up into smaller pieces, more control can be exerted over every piece. This is a very flawed assumption. It leads organizations to design their operations by dividing them up into the silos of departments, functions and too many hierarchical levels. The problem is all these divisions are artificial.

While a segmented approach may have been useful 50 or a 100 years ago, our reality has changed. This is because today there are unprecedented levels of *interdependencies*. This is what creates complexity. We now live in an interconnected, interdependent, increasingly complex world. This requires us to shift our focus and behavior. The good news is that it allows us to operate in a more evolved way. Why? Because in an interconnected world, the levers for success have shifted from competition to cooperation, from win–lose to win–win.

1.2 Useful Lessons from Science: From Parts to Whole

The dichotomy that leaders and managers face today is to lead and manage by "reducing" reality to fragments (silos, functions, etc.) or embrace complexity. This is the same dichotomy that has faced scientists in recent decades. Why does that matter? Because the more we understand nature, the more we are able to learn how to govern our reality effectively. Back in the seventeenth century, thanks to René Descartes, the paradigm was that nature was a kind of clockwork machine. The kind of mechanical laws developed by Isaac Newton were able to explain all phenomena, from the motion of billiard balls to planets. The later discoveries of cells and atoms revealed a more sophisticated reality but one that was still understood mechanistically. Consequently, factories and organizations, from insurance companies to universities, continued to be organized and managed in a mechanistic and divided up way.

Gradually, a shift in science toward a *systemic* understanding as opposed to a mechanistic understanding began to accelerate in the twentieth century. Biologists started to understand living organisms as integrated wholes as opposed to a sum of their parts and physicists began to understand nature through the lens of quantum mechanics where at subatomic level Newton's laws no longer applied.

An increasing awareness of nature based on systems theory has progressively filtered through. We are beginning to understand phenomena within a larger context of interrelations, and to accept the reality of *emergent phenomena*, meaning that there are properties that emerge when parts of a system interact that are not present at the level of the single parts. Every level of complexity gives rise to new emergent properties. This is the sense behind the expression that the whole is greater than the sum of its parts. When we acknowledge this new interpretation of reality, we inevitably see the need for an appropriate way of thinking, organizing, and acting. This cannot be the same way of leading and managing that was appropriate for a mechanistic world.

1.3 The Organization: A Whole That Is Much More Than the Sum of Its Parts

What leaders and managers need to learn from science is that organizations are *inherently systems*. This is the new theory and knowledge that allows organizations to thrive in our contemporary world. Theory, we must state, is not the opposite of practice. A solid theory is what informs effective practice and allows it to be performed in a predictable and repeatable way. Theory is what allows us to progress reliably, beyond the hit and miss of trial and error.

We use the term *system* to mean an integrated whole and the properties of this whole derive from the interrelations among its parts. This understanding applied to organizations was already clearly demonstrated by W. Edwards Deming, the founding father of the quality movement, in the 1950s in his famous sketch of "Production Viewed as a System." The following diagram takes inspiration from Deming's sketch. Our version shows how a *whole organization* can be viewed as a system (Figure 1.1).

We now have a much deeper understanding, thanks to the systems view of nature, of how all systems behave and evolve, and that includes organizations. Understanding phenomena as a system means understanding the context and the relationships involved. It shows that it is not only futile, but also meaningless to try and understand the whole by breaking it down into parts. Putting an organization into context means understanding and managing it *as a whole* in relation to its suppliers and customers and *all* its stakeholders. An organization is intrinsically a system and it cannot be controlled better by dissecting it up into separate parts. An organization, therefore, needs to be understood in terms of the *interdependencies* that exist, internally and externally. Furthermore, a systems understanding of organizations teaches us that at every level of complexity there are "emergent properties." This means properties that do not exist within separate parts but that emerge within the system as a whole, from the interactions and interdependencies in the

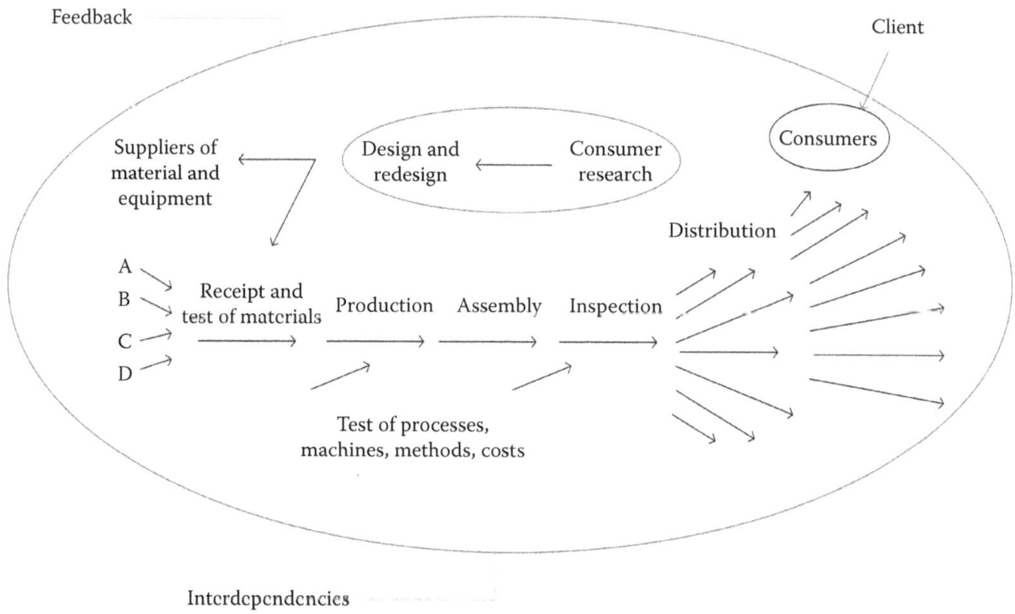

Figure 1.1 The organization as a system.

system. It is the pattern of relationships and interactions in an organization that define its essence.

What we have to learn from science is that by seeing an organization as a system, we are not artificially imposing onto it a structure, the way that functions and silos are an artificial imposition. You cannot treat a living organism as if it were made up of mechanical nuts and bolts. Instead, we need to acknowledge the intrinsic nature of organizations and manage them accordingly. The role of leadership for complexity then becomes very clear; to guide how the network of interdependencies occurs so that the organization can achieve its maximum potential.

1.4 New Leadership

> …the job of a leader is to accomplish transformation of his organization.
>
> **W. Edwards Deming**
> *The New Economics*

In this book, we are working toward much more than improvement. We are highlighting a path for transformation. To transform what? The prevailing silo-based management style into one of system optimization. This transformation requires relentless effort, but ultimately, it allows organizations to consistently fulfill their role in the world with the highest quality and speed. This requires knowledge, method, and tools, and it requires a special kind of leader. We are going to take a look at the qualities and characteristics of the kind of leader who can drive the transformation. Later in the book, we will look at the systemic tools that every leader needs.

1.5 What Is a Leader?

When we understand that an organization is a system, a network of interdependent components, there has to be also a shift in our understanding of leadership as a role and as a practice. Clearly, the old-fashioned command and control style of leadership is no longer valid. Who do we think of when we hear the word *leader*? Moses was the leader of the Jewish people and so was Mohamed for the Muslims. In much more recent times, everyone would agree that Nelson Mandela was a leader, and so was Mahatma Gandhi. In the more mundane world of sport, it is undeniable that Michael Jordan has been a leader and so was Pele. Some organizations "lead" in their field and so do some editorial ventures. In other words, virtually every human endeavor has someone that stands out from the crowd.

If we were to draw a histogram representing the frequency of quotation in management literature, the word "leadership" would probably have the tallest bar. It is therefore puzzling to note how relatively few names come to mind when we think of "leaders," indeed often the word "leader" is attributed to very debatable people.

1.6 Leadership and Knowledge

A leader in today's complex world cannot be a person who improvises. It is true that today there is a wealth of information readily available about every subject matter. We are swamped with information and tips, but as Deming has pointed out, information is random whereas knowledge is orderly and cumulative. A leader today must, by definition, be somebody who has a theory, who "owns" a body of knowledge, not just information, to back their claim to be able to accomplish a transformation within their span of control. As we mentioned before, theory is not the opposite of practice. Theory is what allows solid practice to happen.

The transformation that we advocate here is one of system optimization as a prerequisite for true and continuous innovation. This shifts us away from any kind of command and control style leadership. Instead, the new whole-system approach to leadership is one where competition is replaced by cooperation, where performance is managed systemically, rather than judged, using appropriate statistical thinking and not assessed deterministically, where teamwork is fostered rather than the ranking of individuals.

Knowledge is mandatory but it is not enough. A leader must be able to get their message across; they must be able to address their people in a way that touches their brains but also their hearts. A leader must also have fortitude; they have to "walk the talk" and be an example. A leader has to have the strength that it takes to accomplish something that probably only they have the vision for, and they must be able to clearly communicate that vision. In order for that vision to happen, a leader must have an action plan, a step-by-step guide that people can understand and execute. These are all necessary attributes for a leader, but they are not sufficient.

1.7 Leadership and Selflessness

We have evolved far beyond a simple situation where people just need to be told what to do. A leader of an interconnected organization must provide their people with a vision of life; their words and actions must elevate people's faith in the possibility to build a better world. Jesse James and John Dillinger captured the imagination of many and, sadly, they had many followers. But they were not leaders because their vision was against any acceptable ethics.

A leader is selfless and seeks no power; on the contrary they see their job as servitude. This is not because they are weak but because they understand their role in the world. A leader, at their very core, is an enabler of people's potential. One of the greatest spiritual leaders of our times, Rabbi Menachem Mendel Schneerson, The Lubavitcher Rebbe, says that a "leader has to be a reflection of our own light back to us, so we may see ourselves anew."

A leader must aim at creating other leaders not followers; they do so by elevating their people's ability to address their inner drives and cater for their primal fears. They seek no control over people's emotions; they empower them, they give them the possibility to be the best they can. A leader understands that we live in a world where interconnections and interdependencies easily go beyond our ability to fully comprehend them. The only meaningful role we can play in this interconnected network is to act as a conduit so our organizations can deliver their purpose.

The transformation we advocate is one where emphasis is placed on helping ventures to fulfill their role and purpose in the world, to pursue the goal they set for themselves, to accomplish the vision that has inspired their creation. Almost invariably, this has to deal with enabling the organization to focus on the key elements that determine success. These key elements for success can be summarized as quality, involvement, and flow.

1.8 Quality, Involvement, Flow

Quality is what allows organizations to take a place of prominence. Neglecting quality is also one of the quickest ways for star companies to fade fast. Sadly, in a silo-dominated world, quality is often relegated to a department.

A leader, instead, sees quality as something larger than life, as an all-encompassing attribute that permeates everything we do. Quality is grounded in the belief that management in any shape or form is prediction. When we are unable to predict, we are unable to know the outcome of any decision.

Flow is the result of our actions; flow of materials, information, and money. Flow is associated with motion and therefore with evolution and the pace of change. It is the derivative of space and time, and it reminds us that no real value can be created unless a certain amount of change is taken into consideration.

Involvement of people is what enables quality and flow. The foundation for any involvement is the teamwork that allows organizational systems to function. We can build a system by asking everyone to subordinate to the goal of that system, to give up on something personally for the greater good. It often works but invariably leads to people working mechanically. Eventually, it takes away some pride in workmanship and replaces innovation with compliance. A leader understands that a system can develop and continually improve its results toward the stated goal if people in the system see that what they do for the system actually enhances their personal life. A leader understands that there is no conflict between self-fulfillment and subordination to the goal because a correctly designed system allows an individual to gain more by subordinating to the goal than they would achieve independently. A leader understands that the more a job reflects who a person really is, the better it is for the individual and the system as a whole.

A leader does not exist in a vacuum; they can only exist if they are surrounded by people who acknowledge them. They select their people and those people recognize their leader as such. A leader must evolve and so must their leadership. While the core tenets of their leadership can be everlasting they, personally, will not be. A leader must be ready to pass on the baton at the right time. A leader is a leader when they lead, but also when they stop doing so.

1.9 Leadership Thinking: Framing the Change as a Conflict

How can leaders and managers avoid the trap of taking actions based on outdated thinking? It requires exercising some brain muscle in order to fend off sclerosis in the way organizations function. Avoiding the risk of stagnation in thinking and action means developing the ability to question the way things are done. Leaders have to be able to take a step back and identify and challenge the assumptions they make. They have to learn to ask the right questions and there is no time to lose.

To understand things systemically, we need to educate our brains to "see" things systemically and to solve problems systemically. There are powerful tools available that help create the synapses for positive change and we dedicate a whole section of this book to the thinking process (TP) tools from the theory of constraints (TOC).

One of those tools, the conflict cloud, teaches us that a highly effective way to frame a complex situation is to think of it in terms of conflicting positions, legitimate needs that provoke those conflicting positions, and a common goal. We can then surface all the assumptions (mental models) that keep us stuck in any given conflict or situation of blockage. Once we have identified all these elements, we have a way to allow a win–win solution to emerge. This remarkable tool can be used at a variety of levels, from solving day-to-day problems to analyzing and solving highly complex issues.

Let us take managing organizations and complexity as an example. Here, we are going to look at the conflict of complexity, in other words, two possible and conflicting approaches to dealing with complexity.

The first position in the conflict is "Manage complexity by breaking it up into its parts, or structures." Why would we want to do that? What is the *need* we try to protect by adopting this position? We can verbalize that need as *understand the components*. The opposite, conflicting position is "Manage complexity by focusing on the interactions and dynamics, or patterns."

Again, why would we want to do that? What is the *need* we try to protect by adopting this position? We can verbalize that need as *understand the interdependencies.* The common goal to both positions is to *manage complexity.*

Now that we have pinned down the conflicting positions, we can start to surface all the assumptions that keep us stuck wondering which of these two positions is the best way ahead.

If the goal is to *manage complexity* and we assume that *an element of complexity is its components*, then that leads us to want to *break the complexity into its parts* to understand them better. On the other hand, if the goal is to *manage complexity* and we assume that *complexity is generated by interdependencies*, then we want to *focus on the interactions and dynamics, or patterns.*

How do we move forward out of this conflict? How can we find a solution that allows us to achieve the common goal, to manage complexity, and at the same time respect the legitimate needs to understand the components as well as the interdependencies? How can we do that without swinging like a pendulum between two opposite and conflicting positions?

This is where we see the power of the conflict cloud. We take the two conflicting positions and we surface the assumptions that keep us stuck. Why do we believe that we have to *either* break the complexity into its parts *or* focus on the interactions and dynamics? It's because we make a series of assumptions, and if we can invalidate those assumptions that are essentially mental models and do not have any value of truth, then we can find a way ahead.

We can verbalize the assumptions between the conflicting positions of "parts" versus "interactions" here as follows: (1) the whole is equal to the sum of its parts; (2) no new properties emerge from interactions among the parts; (3) interactions among the parts are always and only deterministic and linear (mechanistic view).

Now that we have verbalized those assumptions and they are "out in the open," we have cleared a path to find a solution that will allow us to lead and manage organizations in a way that is appropriate, adequate, and effective for today's complex reality.

Chapter 2

Why Are We Stuck? The Inherent Conflict of Organizations

There are three major concerns that every organization today should focus on in order to deliver their goal. These are quality, involvement, and flow. Quality is grounded in the belief that management in any shape or form is prediction, and the ability to predict is the essence of any epistemology (theory of knowledge). Flow, as we said before, is the result of our actions; flow of materials, information, and money. Flow is associated with motion, hence with evolution and the pace of change. It is the derivative of space and time and it reminds us that no real value can be created unless a certain amount of change is taken into consideration. Involvement of people is what enables quality and flow. The foundation for any involvement is the teamwork that underpins the functioning of organizational systems.

All the undesirable effects that organizations suffer on a daily basis demonstrate that the way we currently design and manage organizations is no longer adequate and certainly it is not based on fostering quality, involvement, and flow. And yet we have still not made the shift toward a better solution. There is something that is blocking us in creating an organization that is adequate for the complex reality we now inhabit. In spite of its clear inadequacies, the traditional hierarchical/functional model of organization still prevails. Why?

The concept of hierarchy is deeply embedded in our psyches. As far back as biblical times, Moses was advised by his father-in-law, Jethro, to create a hierarchy to help him deliver his teachings to the population more effectively. There was a clear advantage in delegating responsibility in a hierarchical way and it became a standard way to build an organization.

2.1 What's Wrong with a Hierarchy?

If we've been using hierarchies for so long, why do we need to change? The quickest answer is *artificial barriers, ceilings, and silos.* When organizations are created and organized in a matrix of vertical hierarchy and functions, an inevitable consequence is the existence of a series of "walls" or divisions that inhibit the resources involved in producing an optimal result toward the overall goal. This leads to frustration at every level, including that of personal fulfillment. When the structure interrupts the natural flow of the goals people are trying to achieve, when nonsensical policies and agendas force them to do things they fully perceive to be without meaning, or worse, prevent them from improving their performance, they naturally look elsewhere to express their potential or vent their frustration. This creates a complete (and artificial) split between an individual and their life at work.

Even if management is dedicated to the notion of continuous improvement, even if they are dedicated to achieving the goals that have been set for the organization, the artificial barriers of hierarchy and functions will interfere with the ability to perform. Maximizing quality, involvement, and flow is simply not possible within an organization that is designed according to anachronistic thinking.

We are back to the Newtonian, mechanistic worldview that believes that the results of the whole organization equal the sum of its individual, separate and hierarchical parts. To exacerbate matters, traditional measurements and accounting methods are based on local optima. They cannot cater for a "whole system" perspective and so their use perpetuates a mechanistic and divisive modus operandi.

2.2 The Silo Sickness

What are the symptoms that indicate something is wrong with the hierarchical/functional approach to organizations? We could describe this as "silo sickness." Communication gets slowed down or completely blocked, innovation struggles to emerge through the ranks and bureaucracy, projects are delayed and go over budget, quality suffers, and reaction time to changing market demands is too slow.

Individuals suffer because they do not have the authority to carry out the tasks for which they are responsible. Careers are limited because people work for a boss who has to improve their own performance in a vertical way. This means that competencies, both technical and managerial, do not find a natural way to develop. This creates frustration due to artificial "ceilings."

At the organizational level, departments carry out their work based on the local measurements and therefore local optima, losing sight of the overall goal of the organization. Production planning, for instance, in the steel industry, can be completely uncorrelated with sales. This is because there is no feedback mechanism within the organization (and poor interdependence with customers and suppliers).

Even more dramatically, silo sickness means that the larger implications of cause and effect relationships that exist in organizations are totally disguised. It takes time for the effects of a cause to propagate through a system. People have no means of understanding the implications of their local, siloed decisions on the big picture. Even the heads of functions are blindsided to them.

For example, let's say that the VP of operations decides to cut 5% of maintenance costs. A moment will come when the organization is unable to make a sale because it is unable to deliver what it has promised due to breakdowns, and nobody will ever know that the problem originated in the cutting of maintenance costs.

These are just some of the ailments of a traditional organization. We can summarize the three worst faults in an organization that operates in silos as follows:

- No feedback mechanism
- No visibility of the customer (and suppliers) in the organization
- No learning cycle based on feedback

All too easily in a traditional organization the focus becomes power games and personal success as opposed to achieving the goal of the organization. Its very structure invites us to "step on the heads" of others to "get ahead." Having a career path becomes synonymous with "climbing the corporate ladder." Depending on who is "in charge," talented and able people may remain hidden if they are not functional to the agenda of their boss. This is clearly a tragic waste of human ability and resources. Do we therefore need to throw the hierarchy out? The answer is no. What is needed is a kind of hierarchy that better serves the goal of the organization and the fulfillment of those who work there.

2.3 The Inherent Conflict in Any Organization

In the previous chapter, we framed the problem of complexity as a conflict. Below is an image of that complexity conflict captured as a "conflict cloud" diagram according to TOC (Figure 2.1).

This helps us to focus quickly on the root of the problem. We can use the same approach to look at the problem of hierarchy. What is the prevailing organizational structure and what is wrong with it? The conflict is a blatant one. On the one hand:

> A hierarchy fails to acknowledge three critical aspects of the life of a successful company: interdependencies, feedback cycle, and the customer.
> For these reasons, it is not suitable for sustaining a continuous improvement effort. On the other hand:

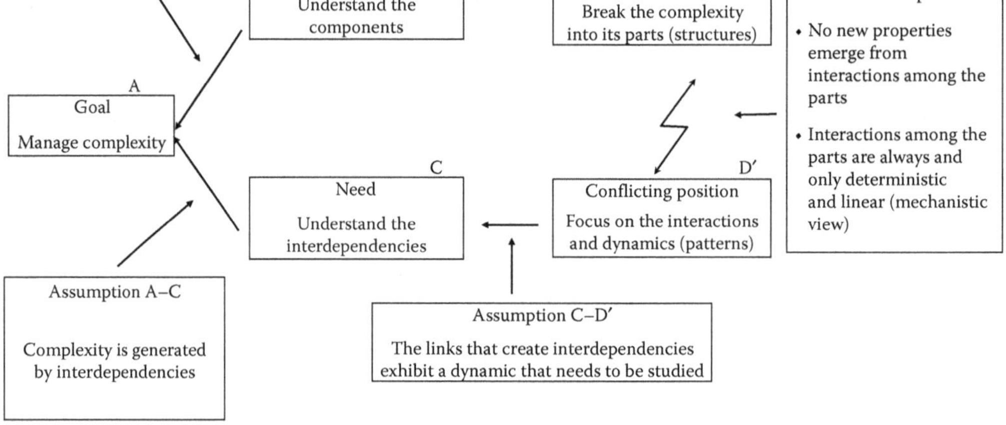

Figure 2.1 The complexity conflict cloud.

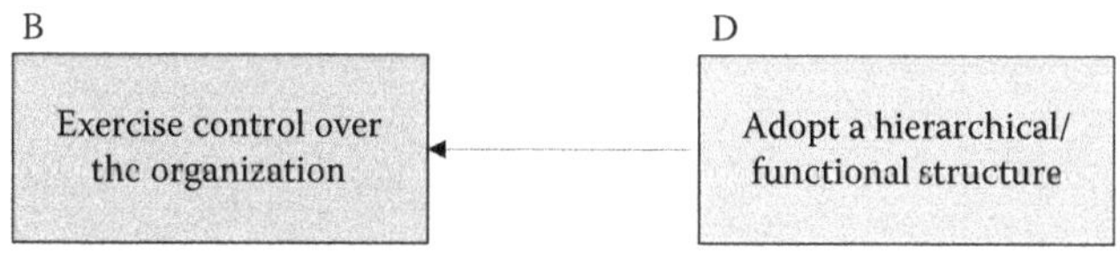

Figure 2.2 B and D in the hierarchy conflict.

A hierarchy seems to cater for accountability and provides a sense of *control.*

The conflict rests, then, on the idea of control and how this idea is translated into a coherent, and coherently measured, organizational system.

Let's reorganize these thoughts using the process of the conflict cloud tool. It is clear that the hierarchical model is inadequate and prevents organizations from unleashing their true potential. The traditional model for *control* is a hierarchical pyramid (Figure 2.2).

The reasons why a hierarchical pyramid exists are the assumptions between B and D (Figure 2.3).

Increasing our capacity to listen to the customer so we can satisfy the needs of the market leads us to NOT adopt a hierarchical model (Figure 2.4). The assumptions between C and D' are connected with what a hierarchical/functional structure prevents us from seeing.

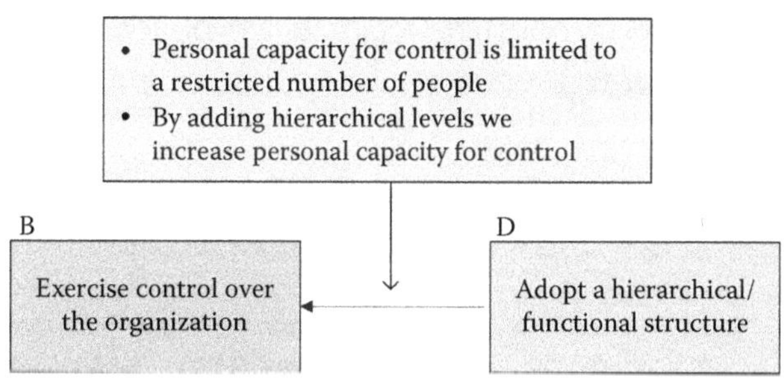

Figure 2.3 The assumptions between B and D.

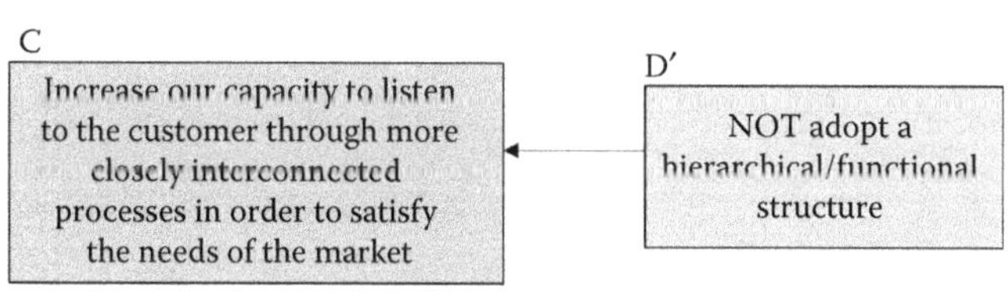

Figure 2.4 C and D′ in the hierarchy conflict.

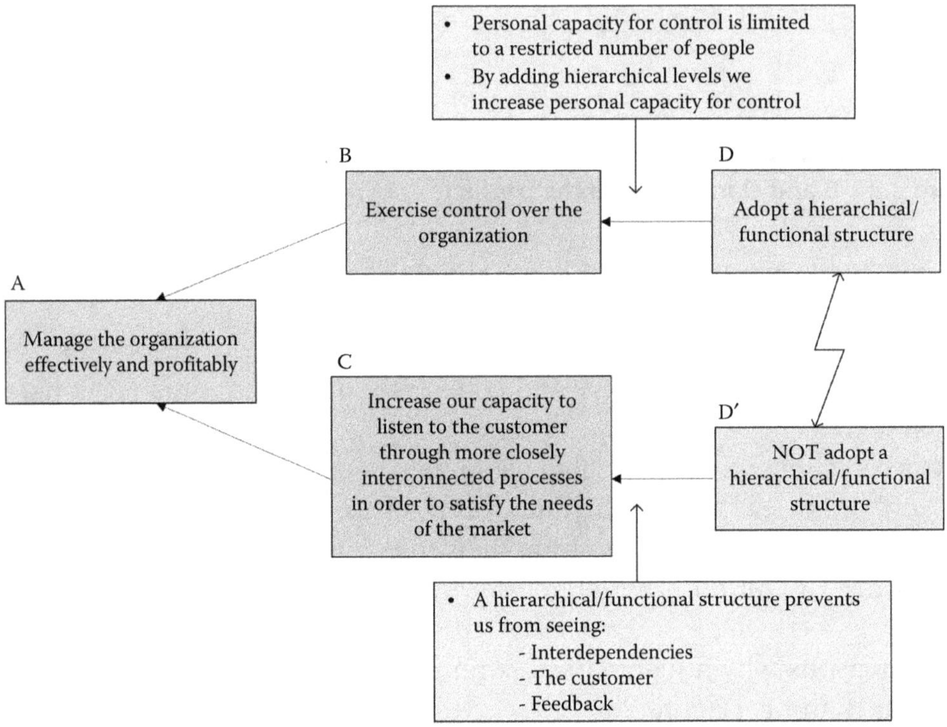

Figure 2.5 The hierarchy conflict with assumptions.

Here are the pieces of the conflict we have looked at put together with the assumptions (Figure 2.5).

The two positions are in conflict because we believe the assumptions in the box on the right (Figure 2.6).

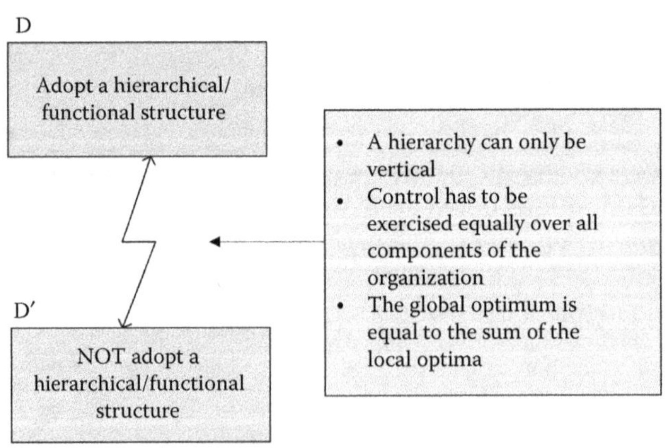

Figure 2.6 The assumptions between D and D′ in the hierarchy conflict.

2.4 Finding a Solution to the Hierarchy Conflict

Challenging the traditional concept of the hierarchy is a step for which many people are ill prepared. What is it about the hierarchy that makes it so untouchable even though it creates so many of the problems that companies experience daily? Clearly, we will not find the answer just by offering the rational explanation of why it does not work. It must be in some doubly wired synaptic connection that is formed very early in our life and continuously reinforced by the environment we live in.

An organization brings individuals with competencies together. By combining these competencies in a suitable manner we achieve the goal of the organization. In other words, individual efforts can lead to a global result if we devise a mechanism that creates orderly coherence in the combination of these efforts. A hierarchy should facilitate the creation of this order. Accordingly, it is *not* the hierarchy that we need to challenge but the kind of *subordination* that a traditional hierarchy calls for. Traditionally, the way we translate the hierarchy into a company structure resembles the well-known pyramid below (Figure 2.7).

As we know, the vertical lines that connect the boxes clearly define the span of control and the accountability of those boxes. Along with the accountability goes the amount of individual power that those boxes have. A hierarchy is then conventionally translated into a mechanism of control and reporting. Accordingly, we can easily conclude that the pyramid cannot achieve the goal of creating the orderly coherence among the individual efforts that we were seeking due to the separations it imposes. Is there another way to create that coherence, hence giving the hierarchy its natural

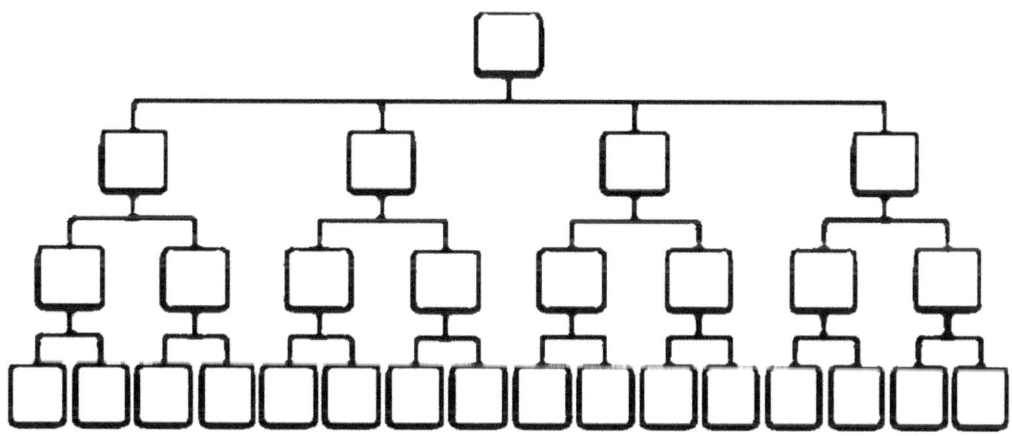

Figure 2.7 The hierarchy pyramid.

role of creating order? There is, but we do need to make a cognitive effort to grasp it. Let's look at that one step at a time.

At its most basic level, if you can imagine an x-ray of a company, any company can be seen as *a set of recurring and nonrecurring activities.* For instance, ordinary maintenance, book closing, production scheduling, shipping, etc. are regularly recurring activities; the introduction of a new technology, the expansion of premises, and the launch of a new product can be seen, in a way, as nonrecurring activities. If we view a company as a system then we must understand how to build the interdependencies that make up this set of activities and how these interdependencies evolve in time. This is an important point: not only do we need to know how these activities are going to take place (this can be done, very straightforwardly, by mapping processes with a deployment flowchart, one that shows who does what, with whom and what is the expected outcome) but also the timing of this evolution and we need a method to control it. As a deployment flowchart can clearly show, virtually any activity that an organization undertakes is *cross-functional*; this realization alone should be enough to dismantle the idea of the validity of a functional organization. Moreover, these interdependencies must be staffed and managed over time (Figure 2.8).

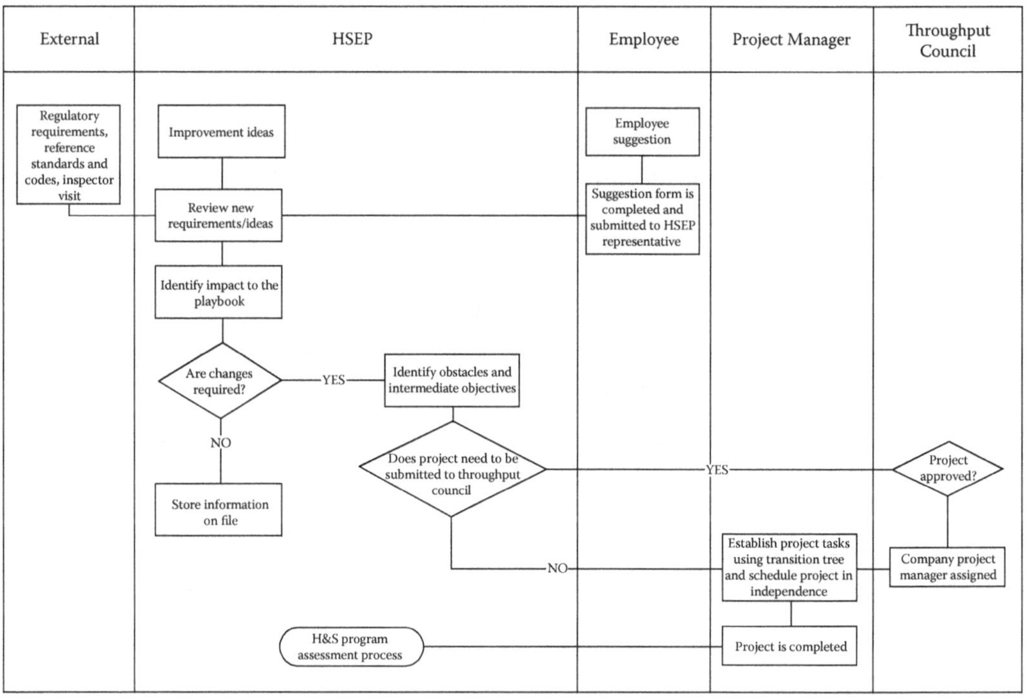

Figure 2.8 An example of a deployment flowchart.

2.5 A New Kind of Organization: Coordination Instead of Functional Reporting

So, to recap, we know that the hierarchical/functional model is inadequate as it creates artificial barriers and does not allow a true understanding of the organization as a series of recurring and nonrecurring activities. An organizational structure should therefore be designed to facilitate the orderly management of sets of activities that are continuously created, coordinated, cross-functional, and that evolve in time. There is a precise name for this in English: projects. A project is exactly this: a network of interdependencies created to achieve a precise goal in a well-defined timeframe. A project is a *system* with a precise duration. A company viewed as a system is therefore a *network of projects*, and the orderly creation and timely completion of these projects should accomplish the stated goal of the network. We are challenging some very basic tenets of the way organizations usually work here, so let's review what are we saying.

1. None of the activities of any company can be performed within the narrow boundaries of a single function; organizations by the nature of their work are *cooperative and cross-functional* and a functional structure is not suitable to support the systemic approach to managing organizations.
2. Any plausible template for an organizational structure that can foster cooperative work must also take into account the way the interdependencies that are needed to accomplish any activity evolve over time.
3. In essence, the management of any organization becomes the management of a network of orderly and evolving-in-time activities. We call them processes and projects and the appropriate hierarchy is exercised through ensuring coordination, not functional reporting.
4. The backbone of any organizational effort becomes, then, the ability to manage the network of processes and projects that any organization is made up of.
5. We can safely say that the springboard to overcome the seemingly untouchable functional structure is the idea of a company seen as a network (with a stated goal) of processes and projects.

A project is exactly this: a network of interdependencies created to achieve a precise goal in a well-defined timeframe. A project is a system with a precise duration. A company viewed as a system is therefore a *network of projects*, and the orderly creation and timely completion of these projects, as well as the management of reliable and repeatable processes, should accomplish the stated goal of the network.

2.6 The Challenge of the Solution to the Hierarchy Conflict

The major problem with replacing a hierarchical mindset lies in the subliminal, unchallenged mental models that make us believe that an organization requires a superimposed control mechanism, be it a boss, a function, or an accounting structure based on cost accounting type considerations. We need, therefore, the kind of thinking processes that help us understand the connections, linkages, and the overall mechanism by which we interpret reality. Reality is shaped in our minds by connections that largely remain unchallenged unless we expose them. By making explicit the cause and effect relationships with which we perceive reality, we have an opportunity to challenge all of those assumptions that limit our ability, for instance, to work within a non-silo infrastructure, that is, within a network of projects. Moreover, in general, an organization where conventional hierarchy has been challenged can function well through enabling a greater ability to "think systems" that is stifled by both current educational systems organized into silos and the corporate world.

It is not enough simply to change processes. Flat organizations could easily turn into a short-lived gimmick unless we systematically challenge the working and, ultimately, existential paradigms governing the image that we have of what it means to live and work together. This requires a considerable cognitive effort and without the right cognitive tools to support that effort, there can be little guarantee of continued success. Later in the book, we will look at the set of Thinking Process tools from TOC that can do precisely that.

There is a price to be paid, however, for growing our intelligence and our freedom. When we remove the barriers, the ceilings, the "stupid" boss who forces us to underachieve, when we take away all this paraphernalia, we are left with ourselves. We are left with the responsibility of no excuses, and the stark and terrifying prospect of truly exploring the boundaries of what we can achieve. What is required is a more evolved form of organization. This organization is neither for the faint-hearted nor for bullies, and it is one that many may continue to reject and undermine.

The good news is that science, civilization, and individuals have evolved beyond the strictures of hierarchy/functions as we understand them today. Contemporary science has revealed the inadequacy of that model. Today we have the science, technology, and know-how to shape organizations into what they intrinsically are—a *system*. We know that every living system is a network, and we can work within those networks to maximize quality, involvement, and flow.

Chapter 3

Where Does the Knowledge to Move Forward Come From? The Giant Contributions of Dr. Deming and Dr. Goldratt

There is no knowledge without theory.

Deming
The New Economics

Experience teaches nothing without theory.

The Deming of American documentary

Information is not knowledge—let's not confuse the two.

Deming
The New Economics

You do not install knowledge.

Deming
Out of the Crisis

We should be guided by theory, not numbers.

Deming
4-day seminars

Without theory we can only copy.

Deming
Out of the Crisis

Where does the knowledge come from to create and manage the kind of systemic organization we now need? Fortunately, in order to build, lead, and manage systemic organizations fit for our complex contemporary world, we do not have to reinvent the wheel. The fundamental elements of knowledge required already exist and much of this knowledge is contained in the work of two of the most influential management thinkers of the twentieth century, Dr. W. Edwards Deming and Dr. Eliyahu Goldratt. When we put their work together into a cogent methodology, what emerges is a powerful approach that provides the principles, method, and tools for systemic management. This combined approach will be covered in detail throughout this book.

Both Deming-and-Goldratt were physicists and both approached management from a systems perspective. Much of their work has been misunderstood and wrongly categorized due to people's tendency to think in silos instead of thinking systemically (looking at the big picture). People often confine Deming's work to "total quality" and continuous improvement. Deming never spoke of "total quality." While he advocated continuous improvement, he did so as part of a vast vision of quality as an overriding philosophy for system optimization. Goldratt's work, thanks to the planetary success of his book *The Goal*, is often filed away as a technique for improving production. TOC, instead, is a full philosophy of systemic management that covers all aspects of business.

The contribution of Dr. Deming and Dr. Goldratt to management is quite monumental. It is beyond the scope of these chapters to either summarize or elucidate upon their work. Our goal is simply to present some of the fundamental tenets of their management philosophy so as to provide the appropriate backcloth for the reasoning that will follow.

Between 1986 and 1992, the world was reveling in deregulation and financial engineering, in other words, the wealth creation hallucinations that years later would take the western world to its knees and destroy millions of lives. In that same period, MIT Press published Dr. Deming's legacy to the world, his two books entitled *Out of the Crisis* and *The New Economics.* Still today, every word of these books is relevant, indeed, much more so than many of the books studied in Business Schools around the world.

In these books, written after 60 years of a very prestigious academic and professional career, Dr. Deming lays the scientific foundation for the transformation of the prevailing style of management into one of whole system optimization. In doing so, Dr. Deming ushers in a new economic paradigm that should shape and inform a new science of management.

3.1 Quality and Involvement through Deming

Deming explained that any organization is a system, that is, a network of interdependent components that share a common goal. If we rephrase this concept with the language of network theory, we can say that any organization can be portrayed as a "directed network," that is, a network with a goal. There are four elements in the Theory of Profound Knowledge (TPK), Deming's legacy. Below, we have added the language of network theory to further illustrate these elements. In order to manage such a network following Dr. Deming's indications we must

1. Understand the dynamics (the emergent properties) generated when the components of the *organizational system* interact
2. Understand the *variation* that affects the links connecting the nodes of the network, in other words, the processes with which the components perform their role in the organization
3. Understand the *psychology* of the individual; the subliminal forces that shape the large majority of people's behavior (and massively impact process variation) as well as the collective psychology of the organization
4. Take an *epistemological* stance in respect with our managerial actions; in other words, operate/act according to a theory so as to have a reasonable conceptual handle on the outcome of our actions (the ability to predict how, within a range of variation, our actions will produce the desired result is based on theory).

Dr. Deming stressed repeatedly that leaders must not necessarily have depth of understanding of these four individual bodies of knowledge: systems theory, theory of variation, psychology, and theory of knowledge. Leaders do not have to be people with a PhD. Rather, leaders should have a good grasp of how the *interdependence* of these elements plays out for the life of the organization. Deming's 14 points, amply illustrated in his book *Out of the Crisis*, are the natural offspring of TPK and address the malaise and basic fallacies that affect so many organizations. Individually, the 14 points point at specific mistakes organizations often make; collectively, they represent a guideline for a cultural transformation. In their original formulation, they are as follows:

1. Create constancy of purpose for improvement of product and service
2. Adopt the new philosophy

3. Cease dependence on inspection to achieve quality
4. End the practice of awarding business on the basis of price tag alone. Instead, minimize total cost by working with a single supplier
5. Improve constantly and forever every process for planning, production and service
6. Institute training on the job
7. Adopt and institute leadership
8. Drive out fear
9. Break down barriers between staff areas
10. Eliminate slogans, exhortations, and targets for the work force
11. Eliminate numerical quotas for the work force and numerical goals for management
12. Remove barriers that rob people of pride of workmanship. Eliminate the annual rating or merit system
13. Institute a vigorous program of education and self-improvement for everyone
14. Put everybody in the company to work to accomplish the transformation

Dr. Deming is unanimously considered the founding father of the Quality movement; Quality for him is something larger than life and permeates every aspect of the work of any organization. Quality, in the most general way, is defined as "on target with lowest possible variation" and this is what TPK aims to achieve. Variation is for Deming (as indeed it is) enemy number one and the ultimate hurdle on the road to sustainable success. A word about variation for those unfamiliar with the term—all natural processes are affected by variation. In nature, no matter how hard we try, there can never be two identical actions that generate exactly the same result. Each of the activities that make up the organization and every individual are affected by variation, and this variation influences the possibility of achieving the goal.

Lack of quality, i.e., excessive, unmanaged and misunderstood variation, from boardroom thinking to shop-floor activities, is what robs people of their pride in workmanship, prevents customer satisfaction, undermines relationships with partners, stifles innovation and creates unhealthy and ethically flawed business environments. Excessive variation in organizations, just like germs that proliferate, is the infection that leads to the demise of any economic activity.

Quality is achieved when people are "involved," and that means intrinsically motivated. Intrinsic motivation in individuals is what sustains efforts over time and is achieved only when we see ourselves reflected in what we do, when we take pride and derive joy from our work. The role of a leader is to ensure that no-one is deprived of that joy because the state of mind created by joy is what enables learning and it provides the drive for the innovation that goes with it.

Flow of ideas, materials and money is the result of this involvement and we shall see how Dr. Goldratt has provided formidable aids to sustain that flow.

3.2 Flow and the Theory of Constraints

> Science is simply the method we use to try and postulate a minimum set of assumptions that can explain, through a straightforward logical derivation, the existence of many phenomena of nature.
>
> **Eliyahu M. Goldratt**

In 1982, Dr. Goldratt wrote *The Goal*, a business novel that describes the predicaments of an ailing factory and the path it adopts to move "out of the crisis" (no connection with Dr. Deming there). The book, originally published by a major U.S. name, sold some 50,000 copies before going out of print. Thanks to the vision and the wisdom of Larry Gadd, founder of North River Press, who bought the rights, the book went back into print and sold more than six million copies in the ensuing decades and was translated into 30 languages. For many years, *The Goal* has been mandatory reading in virtually every business school in the western world and still today sells many thousands of copies per year.

The Goal for Dr. Goldratt has been "a blessing and a curse"; on the one hand it made him instantly famous (and somewhat, and deservedly, rich) on the other hand it boxed TOC into the narrow space of manufacturing. Nothing could be further from the truth.

TOC is a monumental body of knowledge heralded by *The Goal* that Goldratt developed in over 35 years of relentless work all over the world. It is a complete system of thought, elucidated by Dr. Goldratt in more than a dozen books written from 1982 to his demise in 2011. While it is garbed in hard science, which allows fairly straightforward and robust applications to the management of operations, at its core it reflects a very precise philosophical worldview.

A "constraint" for Dr. Goldratt is not a hindrance or a weakness, it is a leverage point; as such, it has enormous transformational power. By acknowledging a limitation, we come to terms with it and use it to further our systemic endeavors toward our goals. Choosing a constraint, deciding what is the "stricture" we want our system to be limited by, is the single most strategic decision we can possibly make. The appropriate use of the "stricture" (a loose translation from the Hebrew word Dr. Goldratt had in mind, rationally or subliminally, while formulating his theory) is what produces the best results because it focuses all our decisions. Once we have a handle on the constraint we can develop a full cause-and-effect map of what it takes to use it at its best and, with that, how we can optimize the functioning of the entire organizational system.

Optimizing the system around the good functioning of the constraint produces the maximum throughput, i.e., the pace at which the system generates units of the goal it pursues. However, for Dr. Goldratt, throughput maximization has a profound spiritual meaning: it allows the organization to become the best possible vehicle for the benefit that it sets out to provide in the world. In the cultural heritage that made it possible for Dr. Goldratt to develop TOC, organizations are conduits, or vessels, their goal is to enable the products they develop to play their fullest role in the world, the faster and more effectively, the better. TOC, then, is all about flow.

When the constraint is "physical" (i.e., a machine, a space, a set of resources, etc.) then the steps in the focusing process to manage a constraint-based system are the following:

1. Identify the constraint—choose it
2. Exploit the constraint—make it work all the time and on the optimal mix
3. Subordinate to the constraint—build the whole network of interdependencies in a way that has the statistical capability to support the optimal functioning of the constraint
4. Elevate—increase the capacity of the constraint but, if it moves, in other words, the chosen constraint as a result of its elevation shifts to another set of resources, then
5. Go back to step 1—start all over again

All through the 1980s, Dr. Goldratt wrote several books to elucidate upon what he had so incisively illustrated in *The Goal* and the financial success of his consulting firm, Creative Output, seemed unstoppable. In those days, the economic cycle was very much "supply side" driven. Companies could not ship fast enough, and any improvement would translate into more revenues. In essence, making the production process "more efficient," faster and with way less inventory at hand, would automatically translate into "more money." TOC, at the time labeled as "synchronous manufacturing," very quickly gained the stamp of an efficiency game.

With the end of the 1980s binge and the sobering up of the consultancy extravaganza, Dr. Goldratt realized that the majority of his consultants and the totality of his customers did not understand the thought process associated with managing the constraint. This seemingly straightforward and perfectly logical system approach to the management of a flow of product as well as information proved to be impossible to grasp for the psyche of the majority of leaders. Despite his tireless efforts to explain plainly and rationally the endless economic ramifications of managing a constraint effectively, Dr. Goldratt's message went largely unheard by the C-suite and Boards alike. Moreover, with globalization and the gradual shift toward a much more "demand-side" economic landscape (simply put, people buy less and are more choosy) the world did not seem to be able to grasp the essentially expansionary nature of TOC, as opposed to the myopic view of seeing it as an efficiency technique.

In the light of the difficulties encountered in transferring a true understanding of the depth and breadth of TOC, Dr. Goldratt's ultimate goal became to "teach people how to think." Over several years, Dr. Goldratt embarked on an intellectual endeavor to produce the "cognitive software" people needed in order to grasp what managing constraints really means. The results of his efforts are the Thinking Process (TP) tools of TOC, first published in the 1994 sequel to *The Goal*, "It's not Luck." We will talk extensively about the tools throughout this book. The TP tools are what enable leaders to develop a focused strategy and then link that strategy in a continuum, seamlessly and flawlessly, to meticulous execution. We will take a closer look at the approaches of both Deming and Dr. Goldratt and how their work can be combined in Chapter 5.

Chapter 4

What Is an Organization? The Fundamental Constituents

What is an organization? That may seem like an odd question, but it is the key to finding a solution for all the issues of management and leadership. When we don't ask this kind of fundamental question, we fall into the trap of making the same assumptions that have always been made. We repeat the errors of the past. We can never overcome the dichotomy between parts and whole that we looked at in the previous chapters. Instead, if we can answer this question accurately, then we can understand both conceptually and on a practical, day-to-day level the most effective way to manage.

Leading and managing in a complex reality requires a much deeper level of understanding than might have been sufficient in the past, what Dr. Deming calls *Profound Knowledge*. This kind of knowledge requires a very different logic and an appropriate set of tools. If we want to get a grip on what is required, then we need to be willing to strip back everything we know about organizations to the most essential constituents and build back up from there.

When we overcome a limited and mechanistic perspective, when we put aside the artificial distinctions and barriers that result from silo thinking, we can take a much deeper look. We can then identify the basic elements that make up the life of an organization. At the most fundamental, elementary particle-like level, what do leaders and managers have to deal with? We can boil it all down to four things:

1. The elements of the system under the responsibility of leadership
2. How these elements interact to produce the desired outcomes

3. The context within which elements and their interactions play out, in other words, the market
4. All the stakeholders—board, investors, employees, society at large

These four constituents form the essential core of leading and managing an organization. In order to cope with the ever-growing complexity of these four constituents, an astonishing number of techniques have been developed to help leaders navigate the agitated waters of the corporate world. Many of these techniques have merit and some of them are certainly more than a palliative.

A technique, however, is only a partial remedy. What is lacking is an approach that (a) helps to understand better each of these four constituents and (b) provides a *unified vision* and a corresponding method to tackle them.

Our effort here is to dig deep into the meaning of each of these four fundamental constituents, to interconnect them in a congruent way (join the dots) and then elevate them to a logical and practical realm where they can be understood and managed cohesively and effectively. When the cosmonauts of the Apollo 13 mission ran into trouble, they were able to complete their voyage successfully because they had the Knowledge (that stems from theory), Understanding (that comes from education) and Tools (acquired through training) to do so. In other words, the content of this book serves not just as a compass but as a complete set of guidelines to navigate the uncharted and uncertain waters of the new economic, commercial and professional paradigm.

4.1 A Closer Look at the Four Essential Constituents of an Organization

Let's look in more detail at the four basic constituents that make up the life of an organization.

1. *The elements of the system under the responsibility of leadership* We can say that the elements of any organization are its company functions. Marketing, sales, production, finance and accounting, human resources, R&D, etc. are the nodes that make up the life of any organizational network. Why do we group things into functions? The reason why company functions were established in the first place is because

the prevailing mindset was one of a traditional hierarchy. They are the fruit of a perfectly legitimate need to create accountability. Functions are the natural consequence of a mechanistic worldview where interactions are believed to be like those of balls on a billiard table: simple and mechanical—there is no awareness that interactions may create any emergent properties.

The major problem with the existence of functions is the behavior that they have engendered. In any conventionally hierarchical organization, functions can very quickly become centers of power accumulation. This can lead to turf wars and this kind of divisiveness is only further fuelled by local measurements of efficiency that do not take into account the big picture. The leaders of functions, whether they like it or not, are naturally poised to compete for resources (budget allocation) and almost invariably positioned for personality clashes.

2. *How these elements interact to produce the desired outcomes* The CEO, ultimately, bears the responsibility toward the board and investors for the company results. No matter the metrics that a company may choose, in order to get these results people have to work together, they have to collaborate. How does this collaboration take place? Within conventional hierarchies people are positioned within company functions and report to their functional head. However, none of these functions is isolated. In order to achieve bottom line results there has to be collaboration among all the functions; so, what we witness daily is the mushrooming of initiatives aimed at promoting "teamwork," often driven by the HR department. At leadership C-suite level, this translates into the CEO having to harmonize the agendas of function heads.
3. *The context within which elements and their interactions play out, in other words, the market* People working together in an organization do not do so in a vacuum and even the most self-centered and inward looking organizations understand that "the market" is the reason why they exist. Within the prevailing paradigm of management, the market is addressed by the marketing function and their responsibility is to devise ways to address the market in a way that allows the sales function to generate revenues. In a conventional silo-style organization, the marketing function crunches the revenue projections that prompt the CEO (and the CFO) to take action.
4. *All the Stakeholders—board, investors, employees, society at large* Many "value creation approaches" that currently inform so much of corporate life are grounded in the belief that the obligations of management and

board are essentially toward the shareholders and the shareholders alone. This is a limited view that does not take into account a more complex reality. Beyond the market, companies, just like individuals, exist in a much broader network of conversations with the environment they are in. Indeed, the more our complex reality bites us the more we witness the development of activities that go under the umbrella of corporate social responsibility: philanthropy, sponsorships, and the like all signal the birth of a new slew of activities that invariably become part of company functions, or may even lead to the creation of new ones. Moreover, with the increase of interactions with the surrounding environment, companies feel the need for some form of compliance to some generally accepted standards; risk management, environmental assessment, SOX, quality assurance, safety, etc. The list of obligations for a CEO seems to be endless.

4.2 New Awareness about Interactions

The fact is that the artificial divisiveness that companies tend to impose is unnecessary, unwieldy and out of synch with how we now understand reality. The elements of a network, their interactions, the common goal they pursue and the context within which they exist do not have to be dealt with in isolation. There is an alternative. A science of leadership exists and we must understand that there are fewer and fewer reasons to ignore it.

There is a way out of the blind alley of organizational divisiveness that stems from artificial silos and it begins with understanding that company functions should be seen in a very different way. They should be perceived as *a repository of competencies or subject matter expertise.* In other words, they contain the know-how needed in order to achieve the tasks at hand.

These tasks, no matter how difficult or complex, require some form of interdependence with other sets of competencies; accordingly, organizational design only makes sense when it concentrates on creating the best, easiest and smoothest way to allow these competencies to interact.

Interact for what? To achieve a common goal; invariably, this goal is connected with satisfying the Market. Only when we truly understand what an organization is can we design one. Organizations are *whole systems.* Organizational design, then, is the science, not the magic, of enabling different competencies to achieve the goals for which they have been brought together. We will begin to look at a complete organizational design for complexity later in this chapter.

4.3 Stakeholders and Corporate Social Responsibility

The more we look in depth at what an organization is, at its most essential level, the more we understand that:

1. Artificial silos and divisions internally can only impede its efforts toward its goal and therefore lead to underperformance. (They prevent involvement and flow and ultimately, quality.)
2. No organization exists in a vacuum; it is part of a much bigger context. Failure to understand the interconnections with the bigger picture is harmful, both for the organization and its environment.

Where does corporate social responsibility (CSR) fit into this picture? It often receives lip service but it is in fact a crucial element, far beyond any single function. What the corporate world needs to understand about complexity is that the *only* form of long-term sustainability lies in taking actions to strengthen the physical and ethical environments where they operate. This may seem like "idealism" but it is in fact "realism," meaning based on what we know about reality today. Any perceived dichotomy between "what is good for shareholders" and "what is good for stakeholders at large" lies in the highly flawed assumption of the zero sum game: if somebody wins somebody else has to lose.

Interconnections, internally and externally, are here to stay and they need to be designed and managed for the common good. CSR must stop being seen as some act of reparation for the damages that "shareholder value-driven activities" inflict on society. CSR must never be considered as "doing good deeds" or "the cost of doing business," or worse, a way to look good in the eyes of the public; CSR must become the philosophical underpinning of any sound, long-term economic decision.

As soon as we open our eyes to this blatant reality, we see that all the compliance activities, far from being a burden, are integral elements of economic sustainability that create real value.

4.4 A New Organizational Design

A new organizational design is needed to support the transformation from the prevailing style of leadership and management to one of whole systems optimization where quality, involvement, and flow can be fostered.

We can thank W. Edwards Deming and Eliyahu M. Goldratt for providing us with the scientific and philosophical knowledge as well as the methods and tools to create this new organization. Their work was not aimed at organizational design *per se.* However, any conscientious attempt to apply the systemic principles that their work has masterfully passed on will inevitably bump up against the walls of restrictions of an incompatible organizational design. You cannot "put new wine into old wineskins."

Over the years of working with a Deming-and-Goldratt approach to management, a congruent systemic design has emerged and we will take a look at that in the next chapter. We call it the "choked tube." Conceptually, and visually, it is achieved by designing a set of interdependencies around a strategically chosen leverage point (the constraint). These interdependencies must be well thought out and statistically stable. A "buffer" is put in place to protect the leverage point from disruption but, most importantly, the buffer serves as a way to monitor and control the *whole system performance* toward the designated goals.

The "choked tube" is, of course, a concept and to make it operational we need to translate it into a "something" that we can lead, govern and manage. When Oded Cohen and Domenico Lepore wrote *Deming-and-Goldratt: The Decalogue*, they did not have a clear pattern for building that "something." Now we do. It is called "The organization as a network of projects." We will return to this organizational design later in the book in Chapter 7.

However, to fully embrace and thrive with this new organizational paradigm, we need a new cognitive lens, a new covenant with the concept of wealth creation and new tools to inspire and guide our people. And this is what this book is about: how to transform, step by step, our organization into a conduit for the good purpose for which it was created.

Chapter 5

How Can We Blend the Deming-and-Goldratt Principles in a Practical Way?

We have established what the fundamental elements of an organization are. We also know we have to work with these elements systemically in order to shift away from the damages of silo thinking and behavior and foster quality, involvement, and flow. The knowledge required for such an organization has been developed already by Deming-and-Goldratt. The question then becomes, how can we blend the Deming-and-Goldratt principles in a practical way?

Dr. Deming never slowed down his efforts to promote better management through better understanding of variation and the way this variation permeates every aspect of our lives. His message was loud and clear: reduce variation, promote statistical predictability, and improve quality.

Like Deming, whom he acknowledges, Goldratt sees the organization as a system. In his book *What Is This Thing Called Theory of Constraints?* Goldratt sketches conceptual patterns from Socrates to modern science. He outlines the steps for identifying what is blocking an organization from growing, how to identify what to change, and how to make the change happen in a pattern of continuous improvement. He introduces the conflict cloud, a powerful logical tool that we have already taken a look at, as a link between the scientific method and what he calls the "intuitive" method.

Both Deming-and-Goldratt were aiming at the same goal. They were both using the same scientific rigor and pursuing "intelligence" as a main driver for the transformation of the management style in the western world.

TOC provides powerful focus for a continuous improvement methodology by focusing attention on the constraint, but it lacks an explicit involvement of statistical methods in managing organizations. Deming's philosophy is fundamental, but it can be very difficult to implement as it lacks the practical TP tools from TOC.

5.1 How a Deming-and-Goldratt Methodology Emerged

Blending the approaches of Deming-and-Goldratt is not an immediately obvious solution. At first, it may seem that there are apparent areas of incompatibility. Goldratt's TOC takes a more reductionist stance by focusing on the few elements that account for the overall performance: the constraints. Deming stresses the importance of understanding and managing the statistical fluctuations of the links that make up the network, whereas Goldratt is preoccupied with building the set of measures, policies, and behaviors that ensure proper focus on the elements that dictate the overall performance of the network.

Until the mid-1990s, the two approaches had always been considered separate, and attempts to engage the "Deming community" with the "Goldratt community" were fraught with difficulty. Dr. Deming dedicated his professional life to promoting better management through better understanding of variation and how this variation permeates every aspect of our lives. As we have already stated, his message was relentless, loud, and clear: reduce variation, promote statistical predictability, and improve quality. Dr. Goldratt, instead, made little explicit reference to variation. He outlined the steps for identifying what is blocking an organization from growing, how to identify what to change, and how to make the change happen in a pattern of continuous improvement. Both Deming-and-Goldratt stressed the centrality of ongoing improvement. Both of them used scientific rigor and the pursuit of intelligence. The focus of their approaches differed, but there was a clear common ground.

In the mid-1990s, Domenico Lepore was asked to develop quality management programs for small-to-medium companies in the Lombardy region of Italy. He was fully versed in Deming's work and shared Deming's background as a physicist. Lepore approached Oded Cohen, one of the world-leading experts in TOC, with his intuition that, not only was there no incompatibility between the approaches of Deming-and-Goldratt, they could in fact be mutually beneficial, each one providing a missing link for the other. Cohen invited Lepore to formalize his intuition in the form of a "conflict cloud." This tool is fundamental for framing a cognitive impasse and developing a breakthrough.

This is achieved, as we have already seen, by systematically surfacing all the assumptions (mental models) that have created the state of impasse. As a result of this effort, Lepore and Cohen were able to verbalize the conflict of "Manage according to the Theory of Profound Knowledge (Deming)" versus "Manage according to the Theory of Constraints (Goldratt)."

Let's take a look at the conflict that Lepore and Cohen worked on. In the conflict cloud diagram, we write the two opposing positions in the two boxes labeled D and D′ to the right of Figure 5.1.

Whereas Deming's TPK is clearly based on statistical predictability, TOC is based on speed. These are the needs that lead to adopting the two seemingly conflicting positions.

In the conflict cloud diagram (Figure 5.2), we write the need that spurs the position D in box B and the need that spurs position D′ in box C.

What is common to both these needs? The common that both Deming-and-Goldratt pursue is a sustainable process of continuous improvement of performances.

We write the goal that is common to both need B and need C in the box labeled A as shown in Figure 5.3.

The reality in which this conflict was rooted becomes clear when we surface all the assumptions. This is the process that we use with the conflict cloud in order to "evaporate" any conflict. We verbalize all the assumptions

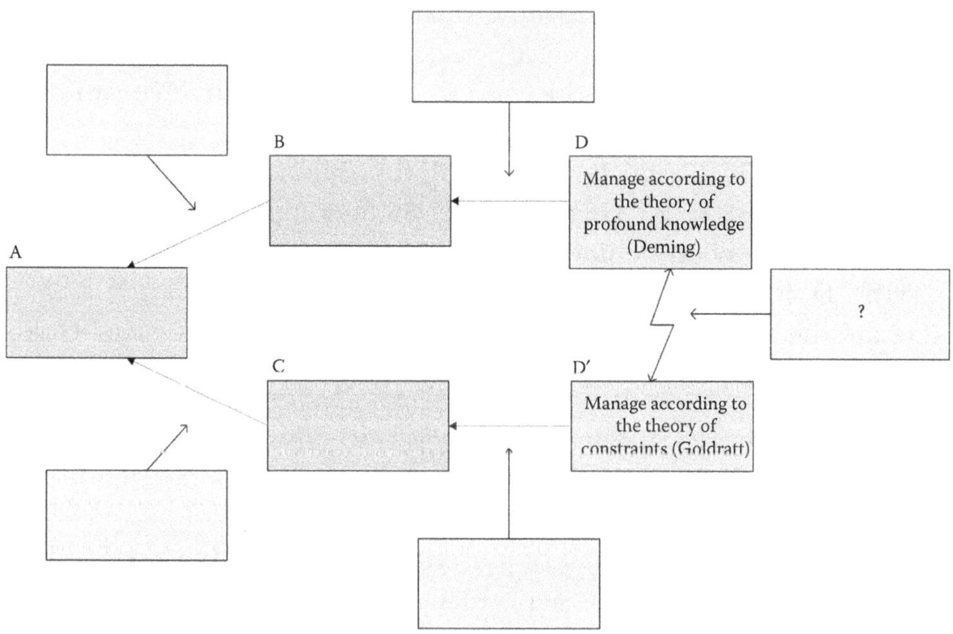

Figure 5.1 Deming-versus-Goldratt conflicting positions.

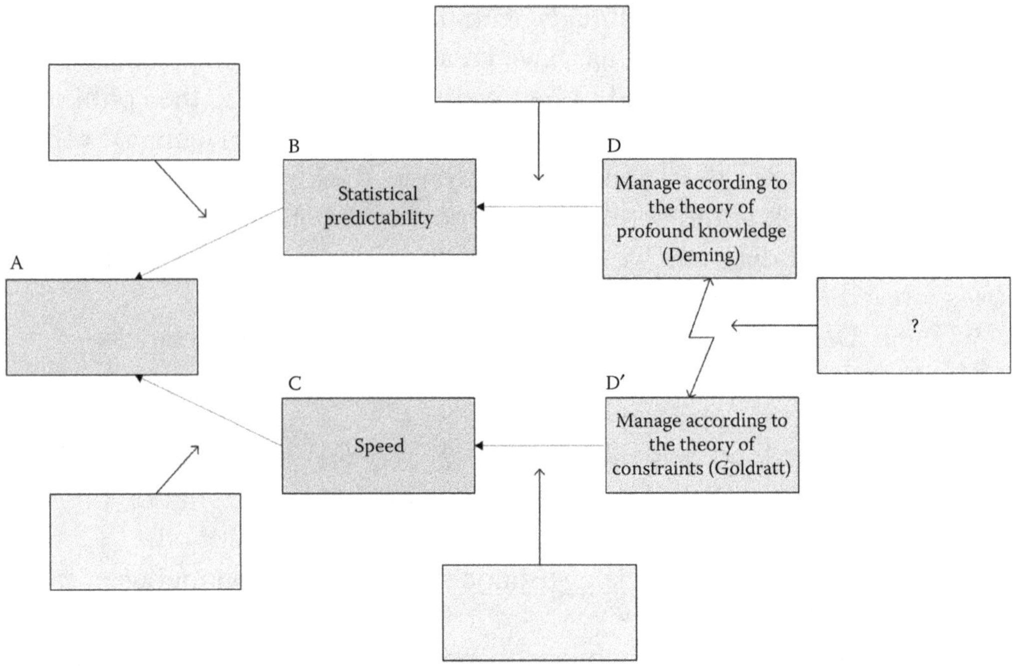

Figure 5.2 Deming-versus-Goldratt conflict needs.

we make that lead us to make the statements in each of the five main boxes, A, B, C, D, and D′. Whereas basic and robust assumptions about reality allow us to live our lives (we "assume" that when we open our front door we can step out onto a hard surface) weaker assumptions are mental models or limiting beliefs that keep us blocked in certain situations. We can derive these assumptions by using the word "because":

IF A (the goal) THEN B (the need behind position D) BECAUSE (assumption(s) A–B); IF B THEN D BECAUSE (assumptions B–D) IF A (the goal) THEN C (the need behind position D′) BECAUSE (assumptions C–D′) IF C THEN D′ BECAUSE (assumptions C–D′).

Here are the assumptions we can surface for the Deming-versus-Goldratt conflict (Figure 5.4):

A–B: The only way to sustain any process is to ensure its predictability

B–D: Deming's philosophy and management approach is designed to ensure (and it is based upon) process stability

A–C: Continuous improvement of performances cannot be separate from the pace at which these performances are achieved

C–D′: The management of finite capacity entailed in the Theory of Constraints maximizes the pace at which units of the goal can be achieved

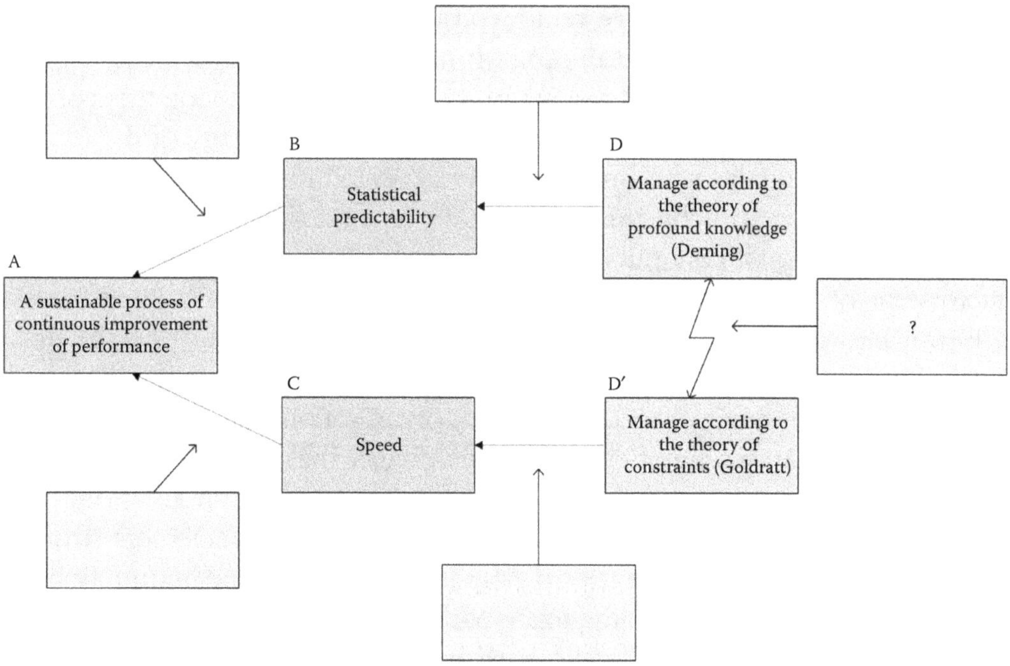

Figure 5.3 Deming-versus-Goldratt common goal.

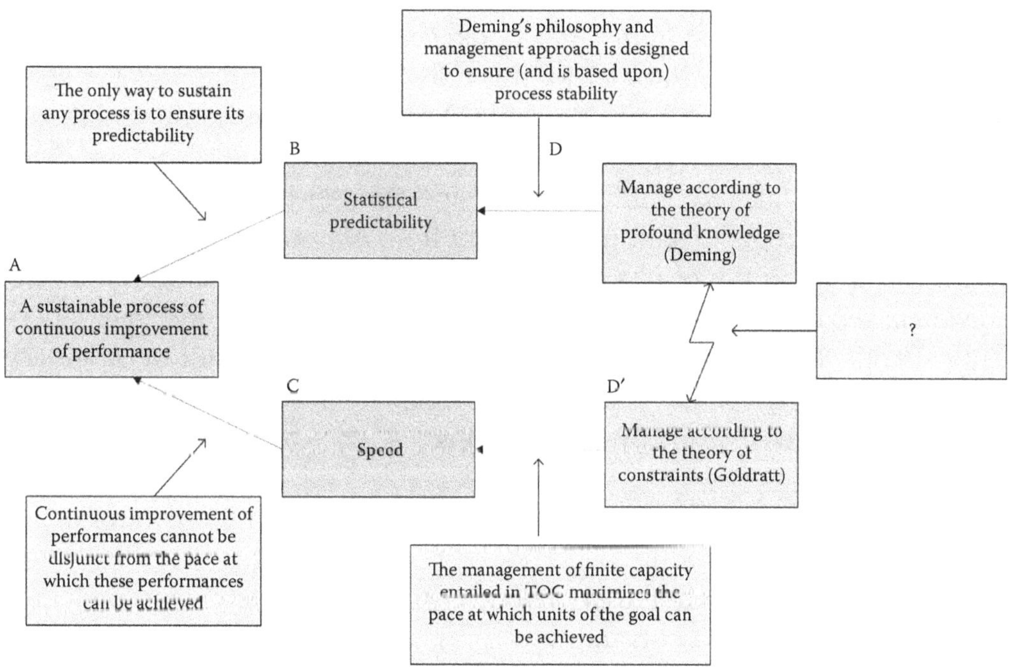

Figure 5.4 Deming-versus-Goldratt conflict assumptions.

The problem is, how can we retain (protect) the need for speed and the need for predictability without ending up in a conflict?

Lepore and Cohen surfaced the major assumptions underlying this conflict, and developed a complete solution that would create a synergy of the two approaches. This analysis became the foundation for a methodology based on managing variation and constraints within an organization with the goal of a sustainable process of continuous improvement. This led to defining the 10 macro steps of the Decalogue methodology, published by North River Press in 1999 (*Deming-and-Goldratt: The Decalogue*). The Decalogue was the first comprehensive attempt to combine cohesively these two allegedly different approaches to the management of complex organizations. The perceived conflict between these two theories mirrors the duality that in science has existed for the last 400 years between a mechanistic approach and a holistic one, and that the awareness of complexity as a challenge for management strongly directs us to resolve. It can be solved only if we challenge the assumptions at the heart of the prevailing organizational paradigm that has dominated the twentieth century. Conceptually, these assumptions belong to the realm of control and measurement and the 10 steps of The Decalogue provide a logical invalidation and a viable replacement for these assumptions; these steps emerge as a practical answer and pave the way for transforming the prevailing style of management in the western world to one of systems optimization, as advocated by Dr. Deming in his 1993 book *The New Economics for Industry, Government, Education*. In order for this transformation to happen it is necessary (a) to redesign the organization and (b) to cater for the learning difficulties that invariably people encounter when the cognitive landscape associated with their work is altered. Philosophically and scientifically, the Decalogue methodology attempts to shift management from a mechanistic, Newtonian worldview in which the results of the whole organization equal the sum of its individual, separate and hierarchical parts, toward a systemic and interdependent network. The shift is achieved by combining the "reductionist" approach of TOC with a systemic view based on interdependencies and interactions. It does so in practical terms by the following:

1. Building interdependent processes managed through the control of variation
2. Subordinating these interdependencies to a strategically chosen element of the system called constraint
3. Designing the organization as a network of interdependent projects with a goal

We will now take a more detailed look at the two main elements that distinguish the Decalogue approach for complexity: managing variation and managing constraints, how they complement each other, and how the Decalogue methodology brings these elements together to manage a network of interdependent projects.

5.2 Understanding the Organization as a System: Flowcharting the Processes

What do we mean when we call an organization a system? Deming's definition in *The New Economics* is the most useful and practical:

A System is a set of interdependent components that work together toward a common goal.

Deming was able to provide a quite revolutionary view of an organization as a system in his diagram from the 1950s that we looked at in Chapter 1. It is a complete shift away from a traditional, hierarchical view of an organization. There are no vertical "lines" to be managed, simply a *flow* of inputs that are transformed into outputs.

In Figure 5.5, Deming includes two essential elements that are completely missing in traditional (hierarchical) organization charts:

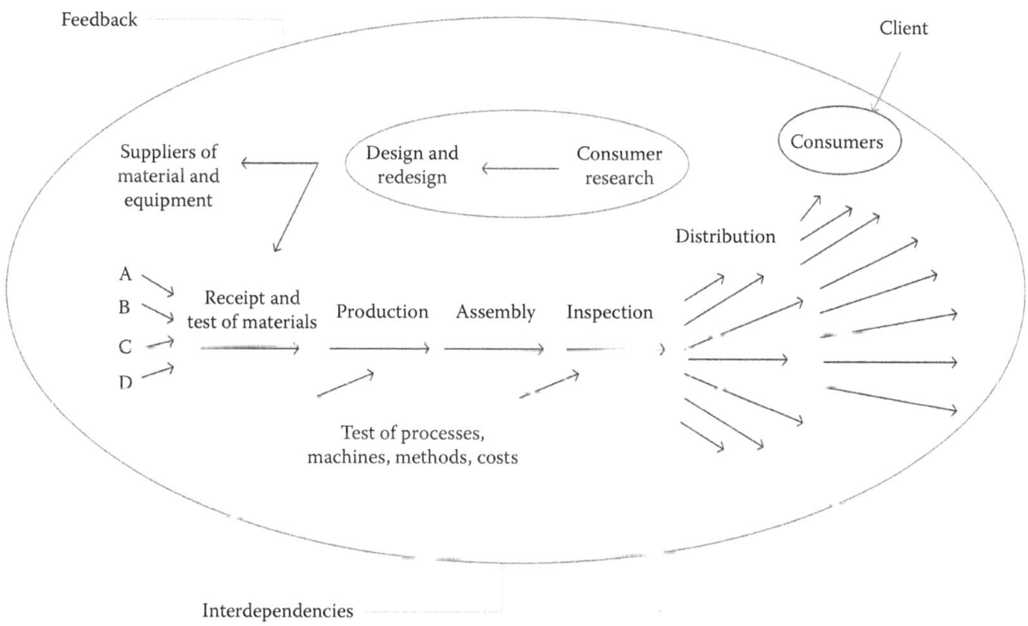

Figure 5.5 Rendering of Deming's diagram "Production viewed as a system."

interdependencies and the connection with the external environment that the organization is part of (customers and suppliers). The modernity and the revolutionary character of Dr. Deming's view are even more evident today, a moment in time when network theory is beginning to clarify the implications of the exponential growth of the interconnectedness of the world we live in.

In order to understand the organization as a system, we have to be able to map out all the interdependencies/linkages that exist. This requires us to understand the company's processes and how to link them together. This can be done simply and effectively by using deployment flowcharts (DFC) to map out every process in the organization, identifying who does what and in what sequence. A DFC describes who does what. It shows the interactions among people in the various phases of the process. It is crucial to know what these interactions are to really understand how the process works and how to improve it.

The steps for drawing a DFC are the following:

1. Identify the boundaries of the process, go through the process following the sequence of events, and achieve an overall vision without too many details
2. Identify the key areas (competencies)
3. Arrange all the stages of the process on a flowchart by competency, according to the sequence in which they are carried out, including all the feedback loops that the conversations entail

When we map out the organization as a system in this way, functional roles disappear and what emerges is a network of conversations that need to take place on a regular basis to make the linkages work toward the goal. This map of processes is not static; it develops over time with the life of the organization. Once we have gained a clear picture of the organization through the mapping process, we identify what we call "key quality characteristics." These are special places inside the organization that we understand have a major impact on the performance of the organization as a whole. Once identified, these characteristics become the key points for where we plan to continuously monitor variation and thus work toward its reduction (if and when possible or appropriate). These actions are the prerequisites for managing variation with statistical methods for continuous improvement. With these two aspects alone, mapping out and managing processes and the variation that affects them, we have the fundamental means for fostering quality and flow across an entire organization. Let's take a look at why understanding and managing variation is so critical for managing organizations as systems.

5.3 Variation and Its Importance for Managing Organizations as Systems

The Decalogue methodology had its beginnings within a Deming perspective, and for Dr. Deming, the very essence of management was "reducing variation." Why is variation so important? Managers need to have the ability to make rational decisions about their organizations, and the work of their organizations consists of processes. Therefore, managers should know how processes behave, both now and in the future. Whether they are aware of it or not, all processes are affected by variation. Indeed, every human process, from getting to work in the morning to carrying out space programs, is affected by variation: a process can never be repeated in a way that is exactly identical. While we can never eliminate variation, statistical methods allow us to understand it, manage it, and take actions to reduce it.

When managers ignore the impact of variation, they do so at great risk as variation strongly impacts the dynamic of the network.

5.4 Variation and Complexity

When investigating complexity, we appreciate even more deeply the need to understand variation. We do not live in isolation; we work and perform activities in connection with other people, usually through cooperation, and these relationships are interwoven in a complex way. It is therefore fundamental to understand that the output of any activity we perform is, in turn, an input for some other activity or person in the network of interdependencies of which we are part.

When we realize the importance and ubiquity of variation, it becomes clear how fundamental it is for a manager, whose job it is to work on the system, to understand the kind of variation the system is affected by so as to make appropriate decisions and take actions that make sense. Not only is it not possible to eliminate variation, it can be passed on from activity to activity, and from process to process. In some ways, dealing with variation is like dealing with infections. If we know what kind of "germ" we are affected by, we can adopt the right countermeasures. We must never forget that infections can spread when nothing is done to limit them.

Let's look at an example of how variation from a supplier's processes can spread to a customer. Suppose we own a factory producing polyethylene film and we buy raw material, in the form of pellets, from a "reliable" supplier, that is, one that we trust. If the density (mass per unit volume) of the raw material is not uniform, then when we come to melt it the result will not be uniform either. The inevitable consequence of this is that our extrusion process will be affected: we may experience a breakdown caused by an obstruction at the extrusion point (a lump of unmelted material), or a rupture of the film (small fragments of unmelted pellets in the extruded film causing non-homogeneity and areas of fragility), or a problem in the cutting phase of the film.

If we do not know the real cause of the disruption, because we do not know the nature and the "amount" of variation that is affecting the system, we run the risk of trying to fix the problem locally (i.e., we fix the problem of obstruction) and then experiencing a different problem immediately after, that is, the film breaks. However, regardless of the nature of the breakdown, we first have to make sure that spare parts are available and that the person in charge of maintenance has the right tools to deal with the problem.

In the end, if we do not have enough information about the nature of variation in the density of the material, we will not be able to make any kind of prediction. Instead, we will inject uncertainty into what kind of spare parts or tools we actually need and, consequently, we will not be able to organize maintenance in an effective way. As we can see from this example, variation has spread out from the supplier to the maintenance process of our factory in a similar way that infections caused by germs have the ability to spread over the body. (We have taken this example from a manufacturing environment, but variation is just as relevant to any environment, from nursing homes to software houses, from advertising to theatre management.) Variation is intrinsic to any human activity, and it strongly impacts the dynamic of the network, because the entropy of a closed system always increases. Variation can be monitored through the use of statistical process control (SPC). SPC, while simple to grasp and use, provides a sophisticated support that helps managers to understand how the system behaves and how their decisions impact the system. (We will look at the tools for this later in the book.)

When we have large, interconnected systems, it is virtually impossible to predict the effect that changing a single component will have on the system overall. We face a problem of complexity, where nonlinear interactions play an important role. Understanding and improving the performance of highly interdependent processes in complex organizations is possible only if we understand the following:

- The variation associated with single processes
- The cause–effect relationships among them
- The impact that processes have individually and cumulatively on the final result

In other words, we have to approach the organization (system) as a whole. It is important to understand that SPC is not simply a technique or a tool for measuring. When we fully understand its importance, then managing variation for the entire, interdependent system becomes a mindset and a way of thinking. If we do not have a clear picture of the behavior of variation inside the organization, and if we allow it to evolve without control, we will inevitably experience an increase in chaos. Furthermore, every time we consider an organization as "segregated" from the rest of the environment (i.e., we do not consider customers and/or suppliers as part of the same picture) we make a grave mistake as we risk transmitting that chaos beyond our own internal system. Keeping variation low is an exercise that has to include all the parts of the system, internal and external.

5.5 Prediction versus Forecast

Why should a leader or manager go to the trouble of understanding the statistical behavior of any process? To understand this fully we have to make a distinction between forecasting and predicting. Managers may be asked to make forecasts, and these are empirical suppositions based on previous evidence and, in many cases, fears and hopes. The usefulness of such forecasts is limited and reliance on them perpetrates a totally inadequate empirical approach to management. Leaders and managers, instead, need to be able to predict how any given process will behave in the future as a result of their decisions. This prediction must be grounded in statistical understanding and the epistemological stance it portrays. Statistical predictability means that the process oscillates "predictably" within its upper and lower limits of variation. This is the only kind of prediction that we can reasonably make and the only one that gives rational value to our decisions (if the process, instead, would not oscillate "predictably" then there is nothing we can say about it that has any scientific validity).

5.6 Adding the Constraint to the Organization Viewed as a System

The concept of constraint is well explained in any book of Analytical Mechanics, and it is the starting point for the description of the dynamics of any physical entity. Organizations are dynamic systems and Dr. Goldratt made a significant contribution to management theory by pointing out that we can also identify the constraint of an entire organization, that is, whatever limits its ability to perform.

The intuition to combine managing variation with managing a constraint came from the realization that constraints, just like variation, are intrinsic in organizations, therefore if the idea of constraint is ignored, then no matter how well we manage variation in the system, we will inevitably find ourselves in a system where there are interacting and constantly shifting constraints. This is a highly undesirable level of complexity, what Brian Joiner refers to as the "whack a mole approach to management."

A methodology that combines variation and constraint, instead, allows an organization to make a strategic choice about which element they wish to be constrained by and to design the system around the constraint while managing the variation of all the processes (creating statistically stable processes) that impact the constraint. Generally speaking, the constraint of an organization is what limits its ability to generate value or, more precisely, units of the goal. Considering what we said earlier about flowcharting, we can consider the constraint of an organization as a very particular kind of key quality characteristic because it is the point that dictates the pace at which the system generates value. This is the reason why it is of critical importance. As with variation, the constraint *cannot be eliminated*; we have to learn how to manage it to the advantage of the whole system, and there is no system without a constraint. In the next section, we take a more detailed look at the advantages of managing an organization around a constraint.

5.7 The Need for a Constraint

Although the word constraint may sound negative, in practice it is not. It turns out that we actually need a constraint and that is very useful. This is true for two precise reasons:

1. We need a constraint to facilitate the synchronization of the different processes in the system (the constraint is the element that dictates the pace at which value is generated)
2. We need a constraint because an "unbalanced" system is more solid and it is more easily managed when the impact of variation is taken into account

In light of the above, we can in fact choose the constraint; we decide where it is possible to conveniently (strategically) position the constraint in order to manage it. Let's see in more detail why an unbalanced system is simpler to manage than a balanced one. The simplest way to exemplify the power that stems from focusing only on one constraint in an organizational system is a production flow. However, this is only ONE example. The production line represented in the following example could just as well represent ALL the processes of an organization, or even the various elements in an entire supply chain. It would be a huge disservice to TOC to consider it as something that is restricted to manufacturing and logistics. Indeed, the following line of reasoning can and has been applied for decades to a myriad of business environments.

A balanced plant or system is a production flow where the process stages (for the sake of simplicity, machines) that generate the finished products all have the same nominal production capacity. The goal of a balanced system is to match the capacity of the various stages so that there is no excess capacity in any part of the system.

Unfortunately, this scenario is quite unrealistic since we are overlooking the impact of variation on the different stages of the chain. At each stage, the amount of production will be affected by variation resulting in the impossibility of the system being "balanced."

Let's look at a simple example.

Let's imagine we have a market demand of 10 units of our product and a production process with five stages, where every stage has the same production average and the same amount of variation as the others. On average, every step of the process can produce, and make available to the next step,

10 semi-finished units per hour, with a performance that varies within a range from 6 to 14 units. Let's simulate the first production cycle (Figure 5.6).

The first stage passes on to the second stage something between 6 and 14 units (let's say 11), the second stage will be able to produce any result between 6 and 14 (let's say 14); similarly, the third stage will generate eight, the fourth stage 12, and the last stage seven.

Given this combination of output, which is perfectly compatible with the machines chosen (by average and variation), how much finished product really comes out at the end of the first production cycle?

The first stage will deliver 11 units to the second stage. The second stage is able to process all 11 units. The third stage will receive the 11 units, but, due to variation, it will only process eight of them and pass them on to the fourth stage. The fourth stage has an instantaneous capacity to process 12 units but, as it only received eight, it will only process those and pass them on to the next stage. The last stage has an instantaneous capacity to process only seven units; consequently, one unit will get left behind. The final result is that, after an input of 11 units, only seven are produced.

It is clear that four units are somehow "lost" inside the production chain as work in process. In this case, three units are stuck between machines 2 and 3 and the other unit between machines 4 and 5.

To summarize, even if the market demand was 10 units, equivalent to our "average capacity," we would not be able to satisfy demand because interdependencies and variation make it impossible to "balance" the production.

Under the initial assumptions, let's see what happens as time goes by. Since variation over time is the same (the range is from 6 to 14), and all the stages have the same average value, the second production cycle will be able to use the semi-finished units that are still in the system.

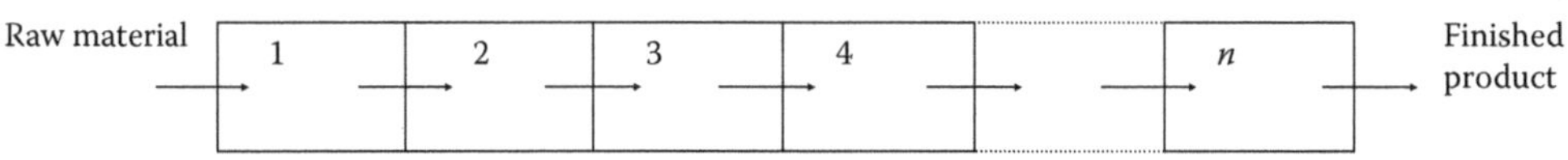

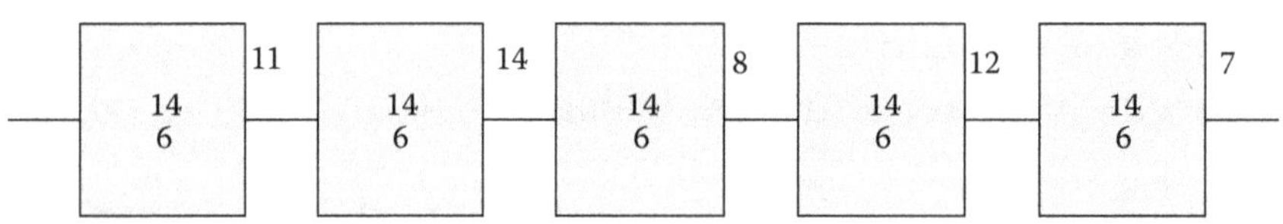

Figure 5.6 First production cycle.

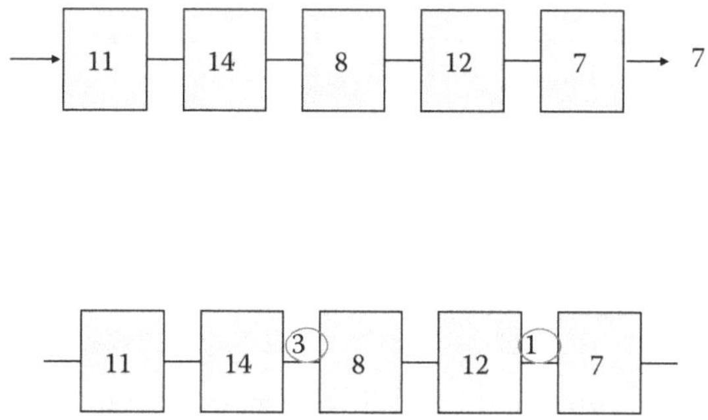

Figure 5.7 Second production cycle.

Let's look at a simulation of the second production cycle (Figure 5.7).

The first stage will deliver nine units to the second stage. The second stage will be able to process all nine units. The third stage will receive the nine units plus the three semi-finished already present in the system and will process all of them and pass them on to the fourth stage. The fourth stage is able to process nine units, leaving behind three semi-finished units. The last stage will use these nine units, plus the one already present as WIP, ending the cycle with 10 units produced and three left in front of the penultimate processing stage.

As the system is made up of a series of interdependent bottlenecks, in every single cycle the instantaneous production capacity of the entire production flow will be equal to the capacity of the stage that has produced the minimum capacity. Moreover, as every machine has a capacity that varies above and below 10 units per cycle, it is almost certain that at least one of the machines has an instantaneous capacity that is less than the average value. Therefore, the real capacity of the entire system will always be something less than 10 units per cycle, at least as long as the stages are interdependent. We are in a paradoxical situation where we produce WIP without managing to satisfy market demand (10 units/cycle). As production continues (the number of cycles increases), there is an increase in the semi-finished material that accumulates between one production stage and another; this is due to the effect of variation in each stage and goes on until the production stages become decoupled (Figure 5.8). Now no stage is forced to remain inactive and each stage can operate independently from the previous one, with a global capacity that is close to its own aver age value.

The price of decoupling is an increase in WIP (Inventory); this means keeping money frozen within the system. With a balanced process we can

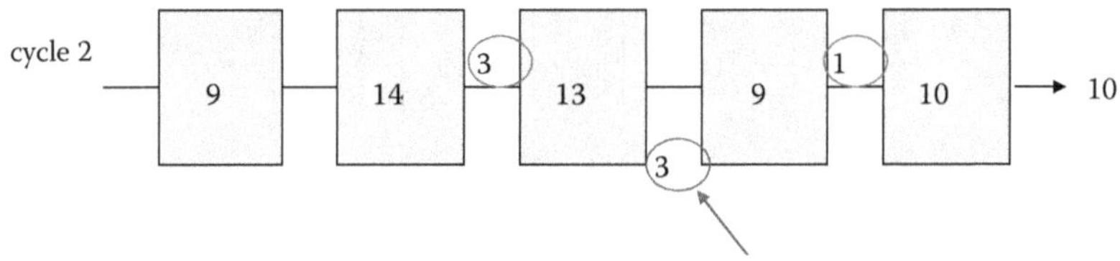

Figure 5.8 Production stages become decoupled.

deliver 100% of market demand only if these two conditions are met at the same time:

- We accept to size the process so that the market demand for every cycle is considerably less than the average value of the production capacity for each stage
- We wait for the condition of decoupling of the production stages to happen

These conditions allow us to deliver 100% of the demand, without generating too much WIP, only by having a much greater production capacity in every stage compared to actual demand. In our example, the system would be able to deliver 100% of market demand if this demand was 8–9 units/cycle (considerably lower than the average of 10 units/cycle).

A different approach is to increase the capacity of some of the stages of the production process so that there is always only one bottleneck in the flow of material (Figure 5.9).

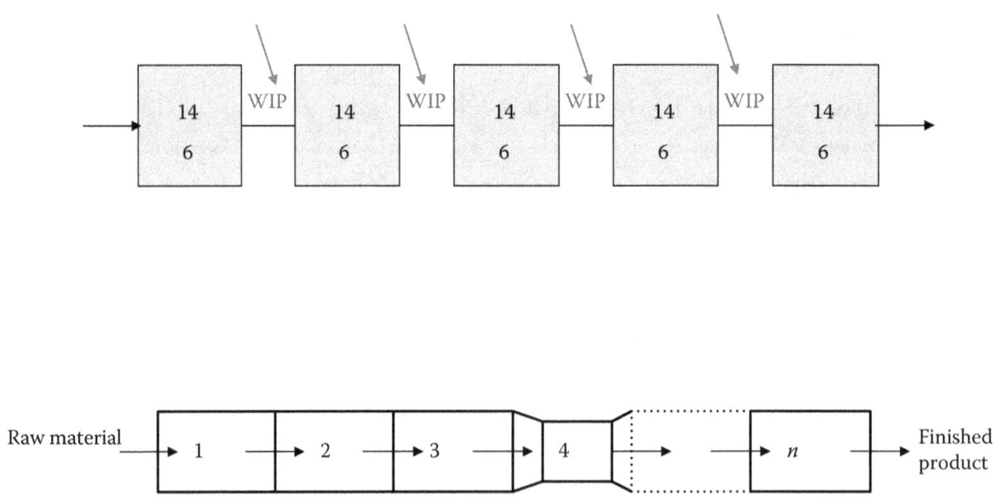

Figure 5.9 Creating one bottleneck in the production flow.

This choice allows us to eliminate most of the WIP, as long as the capacity of the constraint (and therefore of the other stages) is sized on the basis of market demand (market demand must be equal to the average production capacity of the constraint). Also, the flow of the material has to be managed around the constraint; this strategy also foresees an over-sizing of production capacity, but it also has a number of advantages.

Instead of balancing the system, and then trying to improve it by balancing it and reducing variation in all the production stages, we apply the TOC solution. After identifying (or choosing) the constraint, we manage the system around the constraint itself. With this solution, we "unbalance" the system because it is simpler and cheaper to manage.

In an unbalanced system everything revolves around the constraint phase and there are many advantages to this. A detailed plan is made for this phase only. This schedule allows us to manage the whole production system. Also, reducing (global) variation in an unbalanced system means concentrating on and investing in the constraint phase only, not in every single part of the production process. Consequently, increasing the productivity and improving the performance of an unbalanced plant is considerably cheaper and less wasteful in terms of time and energy.

The algorithm that embodies this approach to the synchronized scheduling of finite capacity, which has been largely used by a multiplicity of industries all around the world for more than 30 years, is called "drum–buffer–rope" (DBR); it is a three-step process:

- We identify/choose the constraint
- We exploit the constraint
- We subordinate the system to the constraint

To be more specific, first we identify the constraint in the cycle, and then we decide to manage it effectively. Managing the constraint effectively means managing it so it never stops: in other words, it must never be "starved" of input. In TOC, this is called exploiting the constraint. For this exploitation to happen, every other stage/process in the system has to subordinate. This means that there must be enough production capacity upstream to always produce semi-finished units to feed the constraint continuously and enough production capacity downstream to guarantee that what the constraint produces can be processed. Statistically speaking, this means that the "ranges of variation" upstream and downstream of the constraint phase must have a lower limit of oscillation that is greater than the upper limit of the constraint phase.

As far as WIP is concerned, while in the case of the balanced system it must be available in front of all of the production stages so that uncoupling can occur, in the case of the system with a constraint, WIP must only accumulate in front of the constraint stage to keep it continuously fed. What is more, if the material is released to the whole plant at the pace at which the constraint can process it, the amount of WIP needed to achieve optimal performance can also be kept to the minimum.

5.8 Protecting the Constraint: The "Buffer"

The algorithm we have just described has the aim of maximizing the performance of the system and it is totally dependent on the correct and continuous functioning of the constraint. Managing the constraint effectively implies that it must never "starve" and that is why we must have some kind of protection mechanism in place. In TOC, the protection mechanism is called "buffer management." In a system that is unbalanced around the constraint and subordinated to it, the buffer protects the constraint from variation. The unit of measurement of the buffer is time (Figure 5.10).

Once we know the amount of variation the constraint is affected by (both in terms of input it receives and its processing time), we can define how much time in advance the material has to be ready in front of it. The final, and most important, step is to manage the buffer. Essentially, we have to have an operational procedure that supports our decision-making process; in order to be sure that the constraint works all the time, we have to guarantee that the buffer is never reduced to zero.

Since we believe that statistical predictability is the foundation of management, we must monitor the consumption of the buffer statistically and act accordingly. If the consumption is "statistically stable" we do not intervene; vice versa, if the consumption is statically unstable, then we must take action.

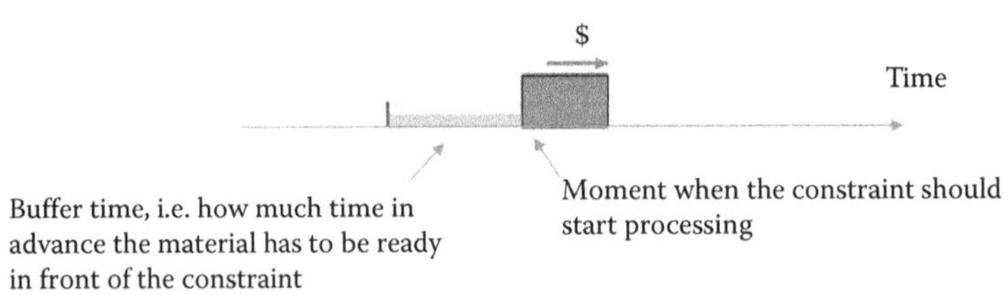

Figure 5.10 Buffer.

5.9 Managing Variation and Constraint(s)

Let's recap on the notion of combining variation and constraint in managing an enterprise as a system. We have depicted our enterprise clearly as a system. We can manage and improve our enterprise by leveraging the insight that comes from a statistical understanding of it. We can manage the variation in the system in the best possible way by increasing reliability, streamlining procedures and improving the quality of every process; we can minimize the impact of market cycles and we can optimize the speed of new product development. Now we need to introduce the idea of constraint, Goldratt's main tenet, into the picture. Goldratt's majestic contribution to management is the understanding that no matter what we do, the speed at which we proceed will be dictated by one, or very few elements of the system. Constraint is a term that may be understood intuitively. However, as it is so fundamental to our approach, let's define it: a constraint is the element that determines how much the system can accomplish. The implication of this definition is that the constraint controls the performance of the system and, hence, this gives the manager responsibility to really control the system. This provided the basis for Goldratt's idea of managing systems through their constraints.

TOC allows us to leverage this reality and use it to increase performance by the following:

- Identifying the constraint
- Exploiting the constraint to make the constraint work at full speed
- Subordinating all the processes of the system to the constraint: build the system around the constraint
- Placing and managing a time buffer in front of the constraint (so "material" will always be available and the constraint is prevented from "starving")

This insight is absent from a purely Deming-based approach. What Goldratt has made clear is that every system has a constraint. If we decide to ignore the idea of constraint, we will soon get ourselves into a situation where the system we operate, no matter how good we are at managing variation, will be made up of interacting and constantly shifting constraints.

Our system is, obviously, a finite one. Its components, as well as the whole system, are finite. This means that we have a choice: we can allow random fluctuations (variation) internal or external to the system to move the constraint erratically, or we can decide strategically by which element of the system we want to be limited (constrained).

In other words, constraint(s), just like variation, are integral to any system.

The idea of process stability combines very naturally and elegantly with the concept of constraint when we place it within the context of the enterprise as a dynamic system. We can have an idea of this from Figure 5.11.

In order to bring together cohesively the idea of constraint with the idea of a statistically controlled system we have to orchestrate statistically stable processes and subordinate them to a well-defined (and VERY stable) part of our system, the constraint. If these statistically stable, well-orchestrated, processes all have the capacity to subordinate to the chosen constraint, then all we need in order to protect the whole system is to place a "buffer" in front of the constraint; needless to say, the oscillation of the buffer must be controlled statistically.

The conceptual elements for a solution to be later described as "choked tube" were first described in 1999 in the book *Deming-and-Goldratt: the Decalogue* written by Domenico Lepore and Oded Cohen (Figure 5.12). It was the first book to be published by Goldratt's exclusive publisher North River Press that was not written by Goldratt himself. The book explains the 10 steps and provides a map to guide and sustain organizations in an ongoing improvement pattern. It has been circulated in tens of thousands of copies, translated into several languages, is recommended reading for university courses internationally and is cited in numerous academic publications.

When we find a solution to a conflict by invalidating the assumptions that lead us to adopt the two conflicting positions, in TOC we call this solution an "injection." The idea of a system that is statistically

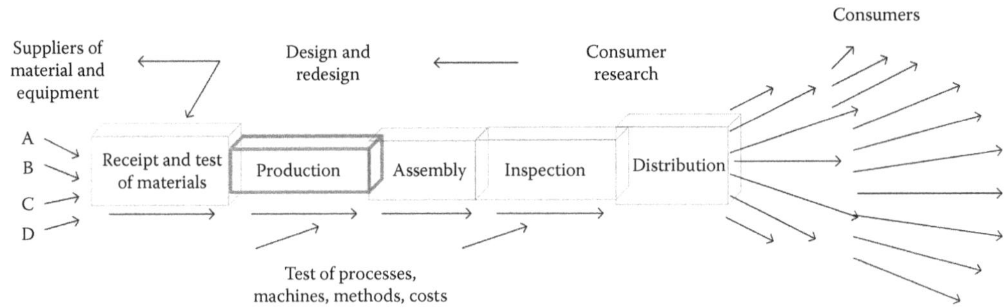

Figure 5.11 Solution of the system choked by a tube.

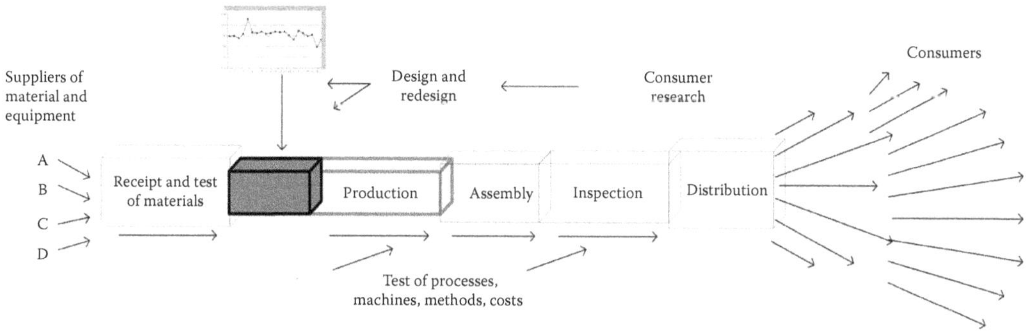

Figure 5.12 System "choked" by a tube with control chart.

predictable and designed around a buffered constraint is a powerful injection and can be very easily explained to anyone with a basic flair for statistics and physics. However, the real issue is, how can we manage such a system?

The answer came in the ensuing years of professional practice following the publication of *Deming-and-Goldratt: the Decalogue*. As the result of a thorough analysis the assumptions behind the Deming-versus-Goldratt conflict became clear, together with the realization of what it would really take to invalidate those limiting beliefs.

5.10 The New Organization: Rolling Out the Solution of the "Choked" System

Compared with the traditional hierarchy, the systemic organization is alarmingly simple. This simplicity derives from the fact that instead of imposing a conceptual model, such as a function, in a systemic organization we reveal the way the organization intrinsically is.

In order to create a systemic organization that combines the two most fundamental elements of a successful system as explained by the theories of Deming-and-Goldratt, we must

- Understand the system we are operating and its intrinsic variation
- Provide a synchronization and protection mechanism that enables its effective management

Let's look again at what that means. Deming's major contribution was to insist on the understanding and management of variation. Every human process is affected by variation because a process can never be repeated in an identical way. Incorrectly managed variation in manufacturing, for example, leads to scrap, waste, and money lost.

It is impossible to eliminate all variation because entropy exists and is intrinsic to any process. However, through statistical methods it is possible to understand variation, measure it, manage it and take actions to reduce it. This requires a mindset of continuous improvement as opposed to monitoring. In spite of the disastrous and costly effects of ignoring this reality, surprisingly few managers are conversant with SPC.

Once we have achieved statistical predictability in our processes, we need to synchronize them and protect them from disruption. This can be done most effectively by identifying the constraint of the system, that is, the element in the system that determines the pace at which the system generates units of the goal. Goldratt's fundamental insight was to understand that we can manage a system by focusing on the constraint, in other words, subordinating the other processes of the system to it to ensure it works to the maximum. We protect the constraint from the impact of variation affecting the other processes by placing a buffer before it. The entire system is scheduled around the constraint using a very precise finite capacity-based algorithm.

All that matters for the success of the organization and those who work within it is speed and reliability. Our goal, therefore, is to create and manage a systemic organization based on process predictability and high synchronization of these processes. The only way to achieve this is to have an organizational structure that is built for and consistent with that very purpose. It is a structure where:

- Interdependencies are clearly laid out through detailed mapping of the processes within the organization
- Variation is understood and managed through relentless application of statistical methods
- A physical constraint has been identified
- A subordination process (to the constraint) is created
- A buffer is placed in front of the constraint

An organization based on the principles of Deming-and-Goldratt is one where the organization is a system designed around a strategically chosen constraint, capable of effective interdependencies internally and externally, and able to innovate continuously.

Deming's system constrained in one point is the injection (solution) that not only unifies the approaches of Deming-and-Goldratt, it also re-defines the ideas of how we do the following:

- Control the system: through the constraint using buffer management and relentless application of statistical methods
- Measure the performance of the system (throughput accounting)
- Design the system for continuous improvement: the "choked" system

The understanding provided by this injection requires solid knowledge in order to be transformed into a reality. This need led to the 10 steps that make up the Decalogue. The Decalogue contains the 10 steps (algorithm) for creating and managing a systemic organization that is Deming's system constrained in one point.

The 10 steps not only allow for the creation of a system constrained in one point, they also follow the pattern of the PDSA (plan, do, study, act) cycle. This cycle firmly embeds the scientific method within the application of the steps. These steps can contribute to increasing manifold the performance and continuous improvement of an organization and we will take a closer look at them in Chapter 8.

The 10 macro steps of The Decalogue are the following:

1. Establish the goal of the system, the units of measurement and the operational measurement (without a common goal there is no system)
2. Understand the system (draw the interdependencies)
3. Make the system stable (understand variation and its impact on the network, make sure that the oscillation of the processes is statistically stable)
4. Build the system around the constraint (subordinate the organization to the constraint, the only part of the system that can never stay idle)
5. Manage the constraint (protect the constraint from the intrinsic variation present inside the system with a "buffer"—buffer management)
6. Reduce variation at (of) the constraint and the main processes (wider variation implies poorer management; low variation improves predictability, reduces inventory and WIP)
7. Create a suitable management/organizational structure (design the network of interdependencies as a network of projects to improve the performance of the system)
8. Eliminate the "external constraint" (sell all the capacity the system has available)

9. Where possible, bring the constraint inside the organization and fix it there (an "internal" constraint is much easier to manage than an external constraint)
10. Create a continuous learning program (motivate people to learn, improve the system through personal improvement)

These 10 steps represent, clearly, more than a simple protocol; they are a *transformational* process. The Decalogue is not for the faint hearted and its adoption requires the willingness to challenge strongly rooted (and often unchallenged) mental models.

Chapter 6

Why Projects Can Become the Backbone of the New Organization

6.1 How People Come to Work Together: Process and Project

As we pointed out in the previous chapters, the aim of this book is to provide leaders (and anyone who has decision-making needs that impact the life of many) with both conceptual backdrop and practical guidance on how to create sustainable prosperity for their organizations. We see the organizational design as one of the main conduits for that prosperity to happen and we believe that a silo-based structure is largely inadequate for that purpose. In order to replace silos effectively, we need to fully understand what makes up the work of any organization and how it produces its hoped for results.

The approaches of Deming-and-Goldratt are fundamental to making this shift away from silos. Deming-and-Goldratt themselves have presented their work in several books and multiple papers and seminars; in our opinion, their theories and findings represent for management what Quantum Mechanics in the 1920s has been for our understanding of the natural world: a paradigm shift. Moreover, albeit with mixed results, several authors over the years have labored to produce further elucidation on some of their topics.

When Lepore and Cohen published the 10 steps of the Decalogue in 1999, it was the first comprehensive attempt to present a cohesive Deming–Goldratt based approach to management, aimed at creating sustainable growth in organizations through a process of ongoing improvement. If we wished to be creative about it, we might consider the Decalogue to be an emergent property of resolving the conflict of Deming "versus" Goldratt. It provided an algorithm to develop the foundation for economic sustainability and wealth creation in any organization and, from a scientific viewpoint, contained all the elements of systems management as advocated by Deming-and-Goldratt. The Decalogue also laid the foundation for understanding conceptually how an organization should be built through the concept of the "The Chocked Tube": a System designed around a physical constraint with a buffer (and its main processes) monitored and understood statistically.

Since the beginning of this century, further theoretical development of the Decalogue, prompted by practical on-the-field work with a myriad of organizations in Europe and North America, has led Lepore and his partners to take another step toward a fuller verbalization of an organizational design coherent with the precepts of the Decalogue. Before looking at that, we need to understand what is a process and what is a project.

As we said before, fundamentally any organization is engaged in two kinds of activities: *repetitive processes* and *one-off projects*. Repetitive processes lay the foundation for the life of the organization. This is what statistical understanding is all about: ensuring the reliability, dependability, and predictability of processes, the very fabric of quality.

One-off projects are what propel the organization to success; new ideas, new technologies, innovation, experimentation are the *raison d'être* for any organization. However, they are only possible and produce results when they emerge from strong roots in quality. Just like trees, the stronger the roots, the healthier the trunk and the branches. Just like trees, only the trunk and the branches are visible but without strong roots the tree would have no life.

6.2 Systemic Project Management

Process and projects, as we said earlier, have very many things in common; they both are, essentially, a set of activities aimed at a very well identified goal. Both have, in general, a timeline for their execution and both need to be operated with a certain degree of statistical predictability in order to be effective (and cost effective). What is the difference, then? Conceptually, not much; organizationally, far too often, the difference is huge.

Even the most siloed, locally minded, and hierarchical organization recognizes the need for a "Quality System" (and the market very often demands its adherence to some international standard). A Quality System (QS) is an accepted and operated set of procedures that allow the members of an organization to build cohesiveness and repeatability in their collaborative efforts toward a designated goal. In an intelligently built QS, these procedures mirror basic processes and these processes are obviously cross-functional. As a result, operating a process, something that is inherently repetitive, becomes an accepted way of functioning for any organization. Moreover, a large chunk of each of these processes takes place within the span of control of one or very few company functions.

Projects, instead, take the predicaments of a siloed structure to a new and different level, and for the following two reasons:

1. A project is, by definition, a one-off activity and, more often than not, draws resources from a variety of company functions, in this way increasing the need for collaboration among people accustomed to working and thinking locally.
2. A project is, more often than not, something that brings innovation, a positive disruption in the life of the organization. Cognitively, a project challenges, albeit in a desirable way, the all too human tendency to resist change in an organization.

A project, in other words, exposes in a blatant and somewhat "violent" way how impossible it is for a conventionally hierarchical structure to accommodate the flow of activities.

In 1997, Dr. Goldratt published his manifesto on a generally unchallenged and largely misunderstood body of managerial knowledge, project management (PM). Once again he chose the format of a business novel and in this book entitled *Critical Chain* he focuses especially on the application of PM to new product development. In *Critical Chain*, Dr. Goldratt goes full force to attack the underlying assumptions on which PM rests and his findings are, as so often was the case with his work, game changing. We shall go back to the details of the Critical Chain approach later on in the book. Here, we need to understand how time, human resources, and money, arguably the life of any organization, can be best combined to achieve a common goal.

A project, Dr. Goldratt argues, draws different competencies together in order to accomplish a specified goal, within a precise timeframe and designated cash outlay. The most strategic use of any set of competencies, and this is arguably the job of a leader, is to deploy those competencies in a way that maximizes their contribution to the overall goal of the organization.

When we optimize based on resources, it very often alters dramatically the logical, sequence-based pattern that leads to project completion. In TOC, this type of scheduling based on the finiteness of resources is called "Critical Chain," as opposed to the well-known "Critical Path."

6.3 The Critical Chain Approach to Project Management

In *Critical Chain*, Goldratt provides a unique and revolutionary insight into this very underdeveloped part of managerial literature. The main tenets of the book are the following:

- Wrong behaviors and mind habits like multitasking and putting issues off until the last minute ("student syndrome") slow down artificially the completion of projects
- Our minds are not trained to assess risks associated with probability distributions
- We must not protect individual tasks but the project as a whole—no milestones
- The traditional Critical Path method for scheduling projects often creates resource contention
- Resolving this resource contention leads to a very different series of dependent events that determine the length of the project—we call it Critical Chain
- It is this chain that we must protect with a project buffer that absorbs the covariance of the project
- Noncritical branches, called feeders, are also protected with a cumulative buffer (not individually) placed at the end of the "feeding chain"
- If we manage several projects in parallel, we must select a finite set of resources called "pacing resources"; they will dictate the pace at which the organization as a whole is capable of achieving its goals

6.4 What Does a Systemic Project Manager Do?

Projects require resources, and these resources have to be managed in order to complete projects on time, within specs and within budget; this is the role of the project manager. However, in spite of the existence of professional bodies, projects are regularly not delivered as planned due to faulty method. On a global basis, this lack of effectiveness in PM is a major drain on resources, not to mention the unnecessary waste.

In an organization that is completely focused around projects, the role of the project manager takes on a whole new dimension. *Every single person in the organization is either a resource on a project or a project manager,* and sometimes both in different projects. PM is therefore a key activity. In a Critical Chain approach, the role of the project manager is both critical and yet *simple.* Their job is to do the following:

- Identify the Critical Chain (various softwares can assist with this process)
- Shorten the Critical Chain (here the difficulty is in challenging the assumptions that we make about the impossibility to replace some of the assigned resources with some idle ones. Naturally, we do not underestimate the skills needed for the task but we firmly believe that training can help elevate people's skills and unleash more necessary resources into the system)
- Assign project buffer and feeding buffers
- Update the project with task execution day by day to spot possible delays to, or earliness of, project completion
- Measure buffer consumption (using control charts) so we know that project execution is predictable or that we have to investigate reasons for anomalous buffer consumption (or increase of buffer in case of systematic early completion)
- Exert a special control on the resources involved in tasks belonging to the Critical Chain and alert the resources next in line in order to capitalize on early task completions

Simple, yes, but what happens when people don't do what they're supposed to do? Once everything is clearly mapped out and everyone knows what they are supposed to do and when to do it, and the project manager knows when to alert people to be ready but the task is not carried out, what is going on?

There can be many reasons why people do not act when they are supposed to act. One reason can be a misalignment of authority and responsibility that prevents someone from taking action. Another can be an unverbalized conflict that keeps a person paralyzed and unable to act. This is where the cognitive elements of systemic management come into play and where the TP tools that we will look at in more detail later are essential for smooth PM.

6.5 Quality, Involvement, and Flow through Project Management

Goldratt's novel *Critical Chain* is very intense, heartfelt and sometimes abrasive. In this deceptively simple and fast-paced book, Dr. Goldratt throws down the gauntlet to academics and industrialists alike on what it takes to use knowledge to achieve results. The Critical Chain method is the offspring of a vision of the world and, too often, the elucidation on this vision has been insufficient compared with the wealth of practical details that Dr. Goldratt's applications often attract from pundits all over the world. Like all of Goldratt's revolutionary contributions to management, Critical Chain has achieved very partial results in industry and none at corporate level.

Looking at Critical Chain as a technique for managing projects means essentially missing the point. The reason why tools like Microsoft Project still dominate the way projects are "managed" after almost two decades of relentless efforts to disseminate Critical Chain is that any attempt to use Critical Chain without embracing a purely systemic view of the organization is doomed to failure.

Critical Chain represents the embodiment of a vision of the organization based on pace of flow, people's involvement, and great emphasis on quality. Quality, involvement, and flow are the basic philosophical pillars of the systemic organization, and Critical Chain can play a much greater role in the building of an intrinsically systemic organization. It can be seen, then, as a highly practical way to implement resource-driven, whole system optimization. This was precisely Dr. Deming's lifelong crusade.

In a siloed organization, one where functions have an allocated budget and a set of performance indicators, these competencies (humans with a set of skills) must be used to boost the functional local optimum. Any attempt to take these competencies away from a function and use them for a "company-wide" project is perceived as a threat to the "success" of that function.

There is an inherent conflict between using resources based on local optimum versus using resources based on global optimum. This conflict exists because it is rooted in the assumptions regarding the most productive way of using resources and, ultimately, it is the result of failing to understand complexity. Solving this conflict can enable a whole range of possibilities for any organization and usher in a new paradigm of cooperative work. Critical Chain can be the engine for that new paradigm.

If we accept the simple notion that resources are there to enable the success of the organization toward clearly specified goals, then all we need is an organizational design that allows just that.

The main drawback of a siloed organization is its intrinsic inwardness. It is unable to keep the needs of the customers in clear sight. Moreover, the lack of explicit (and understood) connections linking the work of each function to a common goal prevents a real cycle of learning from customer feedback.

Put in a different way: silos put blinkers on and hinder the ability to see the system that everyone is part of, thus artificially reducing the flow of ideas and activities toward the goal. A project addresses precisely these issues and PM is about coping effectively with the uncertainty in the flow. Moreover, PM deals explicitly with being on time, in specs, and within budget.

How does TOC help us to deal with projects more effectively? Through a precise process of focusing. Here, we list the five focusing steps as applied to PM:

1. Identify the constraint: time dependence, estimate, and resource availability. (What is the constraint of PM? The longest chain of dependent events, not just in a temporal sense but also considering the use of common resources on which the duration of the project depends—this is what is called the Critical Chain.)
2. Exploit the constraint: eliminate student syndrome and multitasking. (Here, we do everything necessary to ensure that the key tasks are not behind schedule, including detailed planning. We protect the total time of the project and not the execution time of every single activity. To do this we put in the project buffer that protects the project conclusion date from fluctuations along the Critical Chain.)
3. Subordinate to the constraint: task dependency, feeders, and the role of buffers. (Subordinating everything else to the critical chain schedule determines the way we plan the noncritical parts of the project. The noncritical parts are called feeding chains, as they provide outputs that are essential for the integration with the critical chain or that feed a critical chain activity. Subordination means that the feeding activities are completed before the planned start of the

critical activity. This is the cause for the introduction of a feeding buffer.)

4. Elevate the constraint: more resources, better planning, and reduce variation. (This is needed when the total duration of the project planned is too long and unacceptable for the customer. In order to reduce project duration, we have to challenge parts of the project planning. Usually, doubling up of resources on the critical chain tends to have a good result. Other ideas include parallel processing and speeding up of feeding chains.)
5. If the constraint moves, go back to step 1: buffer analysis and inertia. (Buffer management helps the managers involved in the project to anticipate problems before they cause major difficulties. Sorting out the problems ensures the smooth progression of the project while proper analysis of the core problem that endangers the successful completion of the project is carried out. Once a core problem is identified, a full cycle of identification of the solution and implementation of the solution should be employed. Last but not least, inertia must not be allowed to become a system constraint. After completion of the first four focusing steps, people can be complacent and think or feel that everything is ok and this must be checked.)

6.6 The Constraint and Project Management

What is the limiting factor or constraint in PM? In other words, what constrains a project in coming to fruition? If the goal it pursues is a well thought out one, we would happily say that this is its duration. Indeed, the "constraint" of a project *is* its duration. Three elements contribute toward this: the way we estimate task duration, the time dependence of the tasks (their logical sequence) and, of course, how feasible it is for that sequence to exist, meaning, how realistic it is for that sequence to be followed. In other words, resource availability. Once these three factors have been taken into consideration (later in the book we shall exemplify how), we derive what Dr. Goldratt calls "the Critical Chain" of the project: the longest chain of dependent events that takes into account the availability of resources. *The Critical Chain is the constraint of a project.*

When a constraint has been identified, the only thing that makes sense, economically and otherwise, is that we move as quickly as we possibly can so we do not waste any of the constraint's time. In the context of PM, this is the meaning of "exploiting" the constraint. There are two major behavioral hurdles that stand in the way:

1. The all-too-human tendency to procrastinate the execution of what we have been assigned, we call it "student syndrome"
2. Believing that multitasking speeds things up. This is a false assumption induced by a siloed organizational design. Multitasking does not speed anything up, quite the opposite, and we shall see why

How do we make sure that everybody is doing their part in completing the project successfully, that is, on time, in specs and within budget? We need to ensure that we can capitalize on "being early." When people assigned to a task finish ahead of schedule, i.e. they manage to complete their task in less time than anticipated, we have to have an organizational mechanism to take advantage of how task dependency plays out. This is achieved by creating a level of protection against the inevitable variation that affects execution. However, this protection is not applied at single task level; instead, we protect the whole Critical Chain, the constraint, at the end of the Critical Chain with a time buffer that we call the project buffer. By protecting the whole chain with an appropriately sized buffer time, we achieve two major things: we leverage the power of covariance (cumulated variation) to reduce the size of protection (calculated as the square root of the number of the tasks making up the chain) *and* we stay focused on moving the whole chain rather than getting stuck on milestone completion. The same thinking applies to noncritical chains that are the feeders to the Critical Chain. We protect them with a "feeding buffer" so as to allow an "as late as possible start." What is the advantage? Threefold:

1. Much less work in progress to handle
2. Less inventory
3. Better timing for capital deployment

Indeed, we are always trying to make things faster. So, if needed, we can go back to the drawing board for better planning, possible resource additions and, ultimately, better procedures for lower variation in task execution.

Buffers are not just for protection; their most useful role is for monitoring how the overall behavior of the organizational system impacts project execution. If complacency kicks in and we become reluctant to challenge mental models, reality will bite us quickly and it will show up in the pattern of

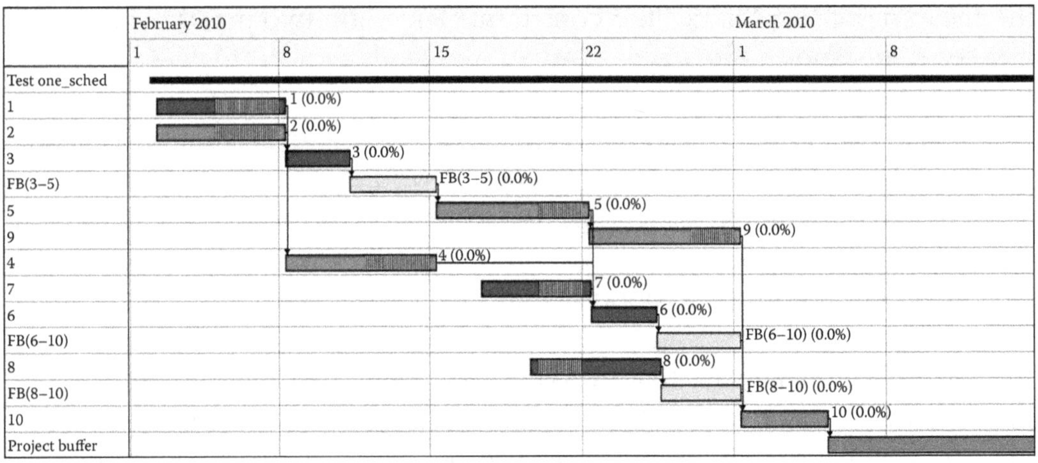

Figure 6.1 Critical chain with feeding buffers and project buffer.

variation (elongation and contraction) of the buffer. Taking action based on buffer variation is the core of a TOC-guided management strategy and it is purposefully called buffer management (Figure 6.1).

6.7 What Does This Mean for a Leader? Toward the Organizational Design of a Network of Projects

In order to accomplish their role, a leader needs an appropriate organizational design. If there are silos standing in the way, what should they be looking at to replace them? In order to see a way out of the conflict between silo and no silo, we need to look at the assumptions in the complexity conflict cloud we referred in Chapter 1. We need to develop an operational solution, what in TOC is called an "injection," that is, a statement that protects both needs of a conflict *and* invalidates the very assumptions the conflict rests upon. We believe that the "chocked tube" design is the conceptual solution to the conflict, so how can we make it operational?

Logically, *if* the essence of an organization are the projects (and processes) it is made up of *and* project buffers offer a very effective real time way to monitor the overall impact of actions taken toward the goals that the organization is pursuing, *then* it makes a lot of sense to consider an organizational design that reflects the very nature of its work. We call this design *the network of projects*.

What we advocate is *the transformation of the conventional silo-based hierarchy into an organization where there is full systemic optimization, made up of projects and processes, where the control mechanism is the management of buffers through statistical understanding.* In this organizational design, company functions are seen exclusively as pools of competencies. These competencies are allocated to projects using a mechanism based on finite capacity, that is, we never schedule a resource simultaneously on two different tasks, we forfeit multitasking in favor of stacking activities, and we foster a culture of promptness as opposed to procrastination (student syndrome).

How can we summarize the major traits of a silo-free organization?

1. Hierarchy is based on the magnitude of the economic or strategic impact of a project
2. Subordination is to the goal of the company, not to a functional boss
3. Doing well locally, i.e. executing proficiently a task, immediately translates into global optimum
4. Individual and collective performances are not assessed but managed
5. The talent and the ambitions of every individual are catered for based on their real inclinations and fear is driven out

In this systemic organization as a network of projects, leaders will be the enablers of meaningful and market-driven projects and will foster a new organizational climate, one of cooperation and win–win.

If the "chocked tube" is the conceptual solution to the inherent conflict and the network of projects is the operational solution, at this point all we need is a protocol, that is, a detailed roadmap for its deployment. In the language of TOC, we need a prerequisite tree. This roadmap is contained in the 10 steps of *The Decalogue* and we believe it is, among other things, a very practical and useful guideline for building a systemic organization (see Chapter 10).

6.8 A New Organizational Design Calls for a New Leadership

The staple of leadership style in a siloed organization is "command and control" in all its different declinations and variations. Command and control, within a conventional hierarchy, leads invariably to some debasement of people's innate desire to learn and participate in something bigger than themselves. It leads to disempowerment.

Why is “command and control,” and the emotional burden that such a style of leadership creates for people, still the prevailing model? It is not because leaders find solace and support in the myriads of ready made, cookbook style, management recipes for success that seem to take up the entirety of the shelves in the business sections of bookstores.

The reality is that “command and control” within hierarchies is easy to understand, regardless of its unlimited fallacies. It is familiar. Changing to something different is a challenge. Indeed, the most fundamental conflict of human existence is the conflict of *change versus don’t change* (we will come back to this in Chapter 9). In order to be tackled successfully, change requires intellectual and emotional willingness. We would argue that leadership and management in this day and age face difficulty in evolving because of the huge cognitive challenge that individuals face in changing their thinking and learning patterns, even their language where necessary. The chapters that follow chart a course of macro and micro actions for systemic change and provide the management tools and, perhaps most importantly, the cognitive and emotional tools for leaders to navigate that change.

Chapter 7

Making the Change Operational: The Network of Projects

In the previous chapters, we have exposed the shortcomings of the traditional hierarchical/functional model and the inevitable damage, both organizational and psychological, that the resulting silos create. In our efforts to lay out a new pattern of thought and action for a systemic alternative we have identified:

- Complexity as the major challenge facing organizations
- The inherent conflict that every organization faces of hierarchy versus no hierarchy
- Where the knowledge we need to solve the conflict comes from (Deming-and-Goldratt)
- The fundamental constituents of any organization
- A practical way to blend the principles of Deming-and-Goldratt through the concept of the "choked tube"
- Projects and processes as the backbone of any organization and Critical Chain PM as the most effective approach for systemic PM

Now, we need to identify an organizational design for complexity that will bring all the essential elements of an organization together and allow them to be managed systemically, in other words, a network of projects, a design that will allow for system optimization and naturally enable quality, involvement, and flow. Before we look at that, we need to understand some fundamentals about networks.

7.1 What Is a Network and How Do Networks Behave?

Once it became clear just how foundational projects (managed through the algorithm of Critical Chain) are for the organizations we have worked with, from metal foundries to nursing homes, it also became clear that network theory was providing important insights for how organizations behave and evolve, and how they can improve. In the previous chapter, we looked at what a project is. Now, we need to know what a network is.

As Dr. Deming taught us, any organization is a network of interdependencies. Network theory is one of the most exciting developments in contemporary science. Our daily activities are now filled with networks. There are various kinds of network in the real world. Some occur in nature, such as a beehive. Other networks are manmade, such as the London Underground (Figure 7.1).

These "simple" networks are not designed with a specific goal and they "evolve" (grow), randomly. The word *random*, here, refers to the way connections distribute themselves inside the network.

In these simple networks each node, "on average," has the same number of connections (links) with a few outliers. The statistical distribution that describes the probability with which these nodes are connected to each other follows a "normal" distribution (Gaussian).

A different kind of network exists where the interconnection among the nodes is greater with some nodes than others. The most connected nodes are called hubs. These networks, known as "scale free" networks, thus have a hierarchy consisting of "visited" hubs and more isolated nodes.

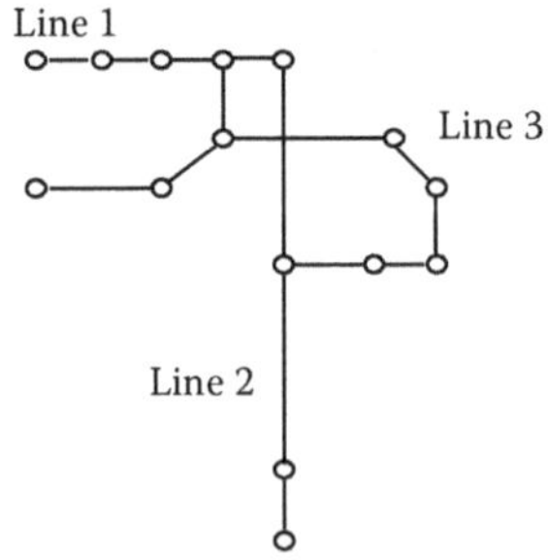

Figure 7.1 Underground network.

The statistical distribution that describes the probability with which these nodes are connected to each other follows a power law. This kind of distribution follows an inverse power relation: for example, a tsunami twice as large as the one that occurred in Asia in 2004 would be four times as rare (less likely to occur) and a tsunami three times as large would be nine times as rare (assuming "power" = 2). When we examine an organization in light of network theory, we are able to unveil the network-like nature of the organization and analyze its behavior and development with a much deeper level of understanding. More importantly, we can consciously design, manage, and operate the organization with a much higher level of optimization.

It is easy to understand that the more nodes and links a network is made of, the higher is the level of complexity. Furthermore, complexity is also amplified by the "interactions" among nodes, which are often nonlinear interactions (interdependencies).

An organization can be viewed as a living, complex network with a precise goal. An important aspect of complexity that we have already mentioned is emergence. This means that a network displays structures and behaviors that are new and that arise from interconnection (they are not properties that belong to the separate elements of the network). Organizations are no different. As complex networks, they also have nodes, links and hubs, and exhibit emergent properties. We know that we can optimize the performance of a system by choosing a constraint and subordinating the rest of the system to the constraint. We can therefore optimize the performance of our organization by designing it as a *network of projects* in which certain resources are the hubs/constraints (or projects/constraints).

By buffering the constraint, we are able to protect it from the accumulation of variation from the processes that feed it. In the same way, we need to protect the hub from variation, and by protecting the hub and subordinating to it, we ensure maximum performance for the entire network. Let's look at this in a little more detail.

7.2 Complex Networks and Real Organizations

How does what we can learn from network theory help us to create and manage real organizations? It does so in a profound way by enriching our understanding and what we do with that understanding. Different types of real organizations are already dynamic systems. We can go further and consider them as complex networks, meaning that they are a structure of interacting nodes and links connected to each other. In these organizations, the greatest undesirable effect (UDE) is "inability to work in a synchronized way." Synchronization implies that every node of the network works toward the goal, and that there is only one goal for the whole network/company.

One way of thinking of a hierarchical organization is that it is made up of clusters where the hubs are not connected to each other. How can we shift this toward a systemic network? We do so by making sure that every element of the network (nodes and hubs) interdepends correctly. How? By making the network evolve into an "oriented network" (oriented toward the goal). This will allow "emergent properties" (a new property that unpredictably emerges from a combination of two simpler constituents) to have a smaller impact on the system, minimizing the (inevitable) peak of variation.

How can we design such a systemic network organization? We can achieve this through the following:

- The sharing of a common goal because this allows the singles nodes to subordinate to that goal
- Creating the right interdependencies among the nodes
- Identifying the right number of links inside the network and putting in place a mechanism to predict how the system will evolve (statistical understanding of variation)

Of crucial importance is the correct identification and verbalization of the goal of an organizational network. This goal will enable the proper identification of the following:

- The logic with which the network has to be designed
- The direction in which the network has to evolve
- Any alternative paths to follow in case of "attacks" to nodes due to the intrinsic variation of its processes (Figure 7.2)

7.3 Connecting "Organizations as Complex Networks" with the Decalogue Methodology

How do we connect the theory and practice of the Decalogue with understanding organizations as complex networks? We do so by recognizing that the UDEs of an organization are in fact emergent properties.

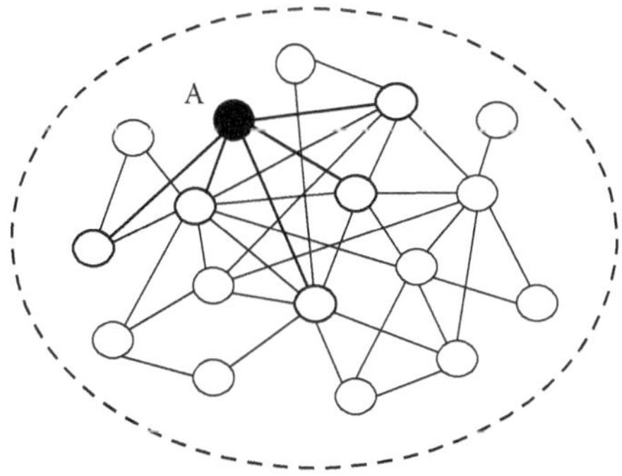

Figure 7.2 Hub in a network.

An "undesirable effect," as the word suggests, is an element of our present reality that we would like to do without. In other words, a UDE is something that "emerges" from the multitude of interactions that exist in the world we live in. Indeed, here we are referring to UDEs that exist in the situation that we are analyzing. By paying attention to the emergent properties of an organization, we can understand a great deal regarding the structure of the network, its interdependencies, and the design of the organization. These properties allow us to identify the following:

- Mistakes in connections and information flow due to an erroneous mapping of the links
- Formation of undesirable clusters that constrain the functioning of the network and prevent it from working in a systemic way

What do we mean by the "nodes" in an organization? They can be employees, machines within a manufacturing environment, or any task that is part of a process within an organizational network. Gathering UDEs, and dealing with them according to TOC, as we shall see later, leads to the choice of a common goal and the correct creation of interdependencies among these nodes, what we call a core conflict cloud.

As a result, we are able to see that the achievement of the goal of the production network is limited by very few nodes that have less capacity than the others. Among such nodes we can identify one, which we call the constraint. This constraint node has to be managed properly and protected from the intrinsic variation of the system.

The constraint is a node whose finite capacity is the most limiting factor for achieving the desired goal of the production network. This node determines the speed at which the network generates products/services and, as a consequence, is strongly connected to the pace of sales.

We can define the constraint as "the node with least capacity that has to be chosen strategically" because it represents the measurement point of the business network. All other nodes are fully linked, within very few degrees of separation, to the constraint node in a way that increases the network cluster coefficient and supports the systemic design.

From a network theory point of view, the constraint is the hub of the network and it is helpful to build the structure of the network itself around the constraint both in order to design properly its mechanics and to model the network in the direction of the goal.

Hubs, the highly connected (few) nodes at the tail of the power law degree distribution, are known to play a key role in keeping complex networks together, thus playing a crucial role in the *robustness* of the network.

To summarize, hubs play the important role of bridging the many small communities of clusters into a single, systemically integrated network.

7.4 Networks, Organizations, and Variation

We have looked at the idea of constraint within the network and true to our approach that embraces both Goldratt and Deming, we must consider variation. The introduction of the concept of variation within a network enables us to look at the process of the output of the whole system and not just at its single nodes. Every process made up of the work of many nodes is affected by an oscillation due to the interaction and the evolution of the network.

Usually, organizations are accustomed to comparing the average output values of their systems as indicators of the system's stability. This forces the decision makers to assess systems based on *incorrect assumptions*. A value below average does not give any indication of the stability of the system because, as per statistical definition, half of the value will be above average on any test, and half below.

The single value, the single sold product, and the single processed item are simply parts of a system whose evolution and reliability are displayed within a process. Therefore, there are fluctuations in processing items, banking services, information given via phone, product assembly, and all other repetitive activities completed through interconnected nodes which also have intrinsic statistical oscillation. How can we know this? Process behavior charts (a better, less intimidating name for control charts, the tool of SPC) display all these local interacting variations of the single nodes as a process and detect the specific limits within which the whole process can oscillate in a predictable way. This can be interpreted as a natural fluctuation of the process in the system. This fluctuation is called variation.

The behavior of processes within a network can become nondeterministic in the case of interacting systems. The consequent emergence of new properties is an important characteristic to be kept under statistical control. At the same time, being aware of intrinsic variation allows us to monitor the development of the network. When we understand this statistically, it helps us predict the emergent properties of the *entire system*.

These two interrelated and cyclical observations are practically implemented by projects developed from the Injections (systemic solutions) derived from building the core conflict of the organization (we will look at this in Chapter 10). Therefore, identifying the constraint and managing the intrinsic variation of the processes occurring in the company by using the methods of SPC enables us to create a systemic network.

7.5 Optimizing Resources through the Organization as a Network of Projects

A network of projects leverages the power of the Critical Chain algorithm developed by Dr. Goldratt to build organizations as systems where control mechanisms, feedback cycles, and rational allocation of finite resources are more effective and in line with the goals of the organization.

Designing and managing an organization as a systemic network allows us to overcome the strictures that interfere with performance resulting from the artificial divisions of a traditional hierarchy/function style of management. The activities carried out in any organization need to be staffed with people with suitable competencies. It will be very cumbersome to deploy these competencies in a timely and systemic way if we organize them functionally. Why? Because any attempt to use a resource allocated to "function A" to perform its competence beyond the boundaries of its function will immediately result in a conflict between the head of the function and whoever has been given the task to deploy that competence, usually a PM. Unfortunately, virtually every minimally complex activity needed by any company to achieve any goal is cross-functional in nature, hence we are stuck in a real dilemma about how to best utilize the resources at hand. The more we realize how paralyzing this dilemma is, the closer we come to understanding the paradigm shift needed to create organizations that are not suboptimized.

The clash between a functional organization and achieving cross-functional goals is, quite plainly and simply, what keeps organizations stuck. This dilemma is the chronic conflict that keeps the science of management from evolving into the real engine of economic growth. Addressing the multilayered issue of how to optimize finite resources to maximize throughput is critical if we are to generate wealth sustainably.

So what happens if we view the recurrent and nonrecurrent activities in an organization as "projects"? Whether we seek to improve the speed at which we manufacture products, install new equipment, organize shipments, or file quarterly closing, we need the *coordinated efforts* of many different competencies. Deploying these competencies in a logical sequence is relatively easy. However, breaking assumptions about the way performances should be controlled and measured seems to be a true cognitive ordeal. The measurement of performances seems to be inextricably connected with a local, in other words, functional, indicator while we all know that what matters is the *global bottom line* of the company. How do we come out of this seemingly irreconcilable conflict? We do so by asking ourselves what company functions are for, and uncovering the obvious truth that functions *should house competencies, not power.*

Engineers, accountants, scientists, and subject matter experts should not be considered members of a "company function." Rather, they should be seen as valuable competencies that can be deployed for the goal of the whole company. These resources, ALL the resources, should be available for whatever "project" the company needs to accomplish.

What we are saying is that *any company should be seen as a network of projects with the global goal of maximizing the Throughput of the company.* The theme of Dr. Goldratt's book *Critical Chain* is that we can maximize the speed of new product development by adopting a particular approach to PM. The implications of that approach are truly far reaching and pave the way for a complete and yet largely unexplored solution to the inherent conflict of any organization (hierarchy vs. no hierarchy).

What we have been able to discover in over 15 years of implementing the Decalogue is that the Critical Chain method that Dr. Goldratt developed can be used not only to maximize the use of the finite resources of a company; this algorithm can be used to redesign the way any company works. Critical Chain becomes, then, much more than simply an algorithm to accelerate project completion; it is the vehicle to integrate, control, and deploy the resources of the organization.

What does the network of projects organization look like? Instead of company functions, there are networks of projects; instead of heads of functions, there are managers of increasingly complex projects that draw their resources from a pool of available competencies with no resource contention; instead of executives that fight for power, there is cooperative work that is in synch with the goal of the company. Instead of often-conflicting local indicators of performance, there is one single driver for everybody.

7.6 A New Kind of Hierarchy

As we have stated, a network is a set of "nodes" connected by "links"; it is not a static entity, indeed it is something that continuously evolves, and its sustainable development is governed by a dynamic dictated by the emergence of a few hubs strongly connected to nodes through links whose behavior is statistically predictable. This dynamic is, especially in the specific case of human networks, nonlinear because of the intrinsic nonlinearity of the interactions between the nodes (people or groups of people). At the most fundamental level, organizations are networks of different communities of practices; it seems only reasonable, then, that the prevailing silo-based, hierarchical/functional organizational design should be replaced with an organic way of structuring such practices.

The network of projects that makes up the life of the organization is a particular kind of network, called directed network, where the "direction" is provided by the goal of the organization. Such networks are called scale free, as we described above, with a hierarchy of hubs and nodes. Accordingly, in an organization there will be *hub-projects*, namely the ones more relevant for the success of the organization and *node-projects*, i.e., projects that are smaller but still necessary for the development of the organizational network. What ensures the connection among these hub and node projects is the finite capacity algorithm for their synchronization fuelled by an appropriate database of resources. Stated simply, two key elements enable the optimal management of finite resources: *predictability* in the execution of activities and their *synchronization*. In other words: IF individual activities are performed with a high degree of reliability, with quality, AND these activities are orchestrated with a powerful algorithm that allows the best possible synchronization toward the stated goal, THEN we have an infrastructure that can maximize the throughput that the organizational system can generate. Within the 10 steps of the Decalogue for managing organizations as systems, based on managing variation and constraints, Step 7 of the Decalogue regarding organizational design becomes the design of a network of projects.

7.7 The Cognitive Challenge

There is a real cognitive ordeal to overcome before we can embrace the idea of a project-based organizational structure: we feel we lose all our familiar reference points. We are accustomed to having "a career" in a functional area with a "boss" that assesses our functional performances and a "bonus" paid to reward a local optimum. This is so true that even if we understand that these things do not make sense we still struggle to give them up. Yes, we may like the idea of a "systemic organization," it makes sense to us, and yet we will balk when someone tries to take away our local certainties: I need a boss to report to, I want to be measured locally (how else?), and I do want my bonus for doing my job well and loyally (to the boss).

Let's see how we can accommodate, at least rationally, for these seemingly indispensable features of our professional life. A systemic organizational structure that is based on process predictability and a high level of synchronization must safely rest on the following:

1. A clearly, indeed super-clearly laid out network of conversations (what everyone needs to say to everyone else to make processes work: input, output, how to measure it and how to improve it; indeed, how the process should work). We can call it "The Playbook."
2. A suitable information system (IS) structure to support these conversations. (We will look at both Playbook and IS systems in more detail later.)

These two issues are neither conceptually nor technologically difficult to address. The network of conversations requires clear ideas and sufficient knowhow on how to operate the company's processes and how to link them together. It can be built in weeks, not months, for any mid-sized company and a few more weeks are needed to expose anyone in the company to the outcome of this work. The suitable IS structure is even simpler: the different pieces that would make up this IS are already available as open source and all it takes is clarity on what an IS should be for. Unfortunately, concepts and technologies are the offspring of paradigms and these paradigms are originated by forces; often we are not trained to understand and control these forces. Let's digress for a second.

7.8 Career Paths and the Systemic Organization

Let's ask ourselves a basic question: why do we choose/accept a certain job rather than another one? Why do we take a position in accounting, production, R&D, marketing, etc. in Company X rather than in Company Y? In the last 20 years of work on both sides of the Atlantic the overwhelming majority of the people we have met would answer: "It would offer me the best chances for a career."

We cannot prove it, but we have strong empirical evidence that jobs are offered (and accepted) by leveraging people's legitimate desire to have a successful career. Even if a functional position will take away most of their professional pride in no time because it forces them to act "unintelligently" and work ludicrous hours for often no reasons other than a boss's agenda, even if it provides them with little learning and, invariably, sooner or later will stifle their possibility to advance, the need to know that "I have a career pattern" is overriding. If this force is so brutally prevailing, how can we accommodate for it in a systemic organization? How can we design a career pattern in a systemic organization built as a network of projects? Let's take another step.

Let's repeat it one more time: the structure of a systemic organization must be based on *predictability and synchronization*. Predictability is ensured by technical/subject matter prowess and synchronization is ensured by the ability to combine these competencies effectively.

A person can have a strong technical competence and love the idea of exercising that competence all day long; another may have a lesser inclination for any particular subject matter but be versed in the art and science of managing complex projects. In a systemic organization, both these people have a chance to progress and this progression doesn't need to be artificially hindered by functional boundaries.

In a systemic organization, anyone, at any time, is part of a project. They lend their competence to a project that is designed, along with all the other projects the company is made up of, to maximize the results of the whole company toward its goal. Every member of staff at any given time is a resource for a project or a PM for a project; sometimes they are one or the other on multiple projects. Some people will develop competencies for managing increasingly complex projects, some others will continue to deepen their competencies and enrich the content base of the company. Most importantly, all of them will naturally be placed within a continuous learning pattern.

7.9 Achieving So Much More

How do we do it? We have learned that the interaction of hubs and nodes gives birth to emergent properties, structures or behaviors that emerge spontaneously from interaction. The years of implementations have shown us that we can consider the Decalogue as an emergent property—something that goes far beyond the simple combining of two separate elements. It is a unique methodology that lays the foundations for continuous improvement and sustainable growth and points at a new organizational model. This organizational model evolves the systemic combination of Deming-and-Goldratt into the design of the network of projects—a vision and a method that can make the systemic organization truly operational. To achieve this, reality requires a higher form of intelligence and ethical awareness. We can accomplish this with the ecology of the mind that systemic thinking provides, and by embracing the values of a world-view where profit and the common good are inextricably interdependent. Both W. Edwards Deming and Eliyahu Goldratt understood that humanity is capable of so much more than we imagine. Whatever situation or business we are in, we can do fundamentally more, and we can achieve that in a meaningful way, not to the detriment of others, but by creating something new, better, and from which all will benefit, from the supplier to the end user. Far from being a utopia, this understanding is rapidly becoming the obvious solution, because it is the only sustainable way ahead.

Chapter 8

Ten Steps for Designing, Measuring, Managing, and Improving the System

8.1 Scope and Purpose of the Decalogue Methodology

Our objective when we adopt a truly systemic approach in an organization is to transform the way the organization pursues its goals to create quality, involvement, and flow. It is not change management, nor re-engineering, nor improving, but *transformation*. The 10 steps of the Decalogue are designed precisely for this. It is not Kaizen, it is not Lean, and it is not business process re-engineering: its goal is not the achievement of better performance because that is a natural by-product; it is a complete paradigm shift in the way the business of the organization is conducted.

The transformational nature of the Decalogue is reflected in its ambition to bring together the elements of the New Economics advocated by Dr. Deming and the largely unexplored power provided by the thinking approach to systems of TOC. Dr. Goldratt has labeled this thinking as "common sense" and called its offspring "thinking tools." After almost 20 years of implementations there seems to be little of common about the "sense" that Dr. Goldratt talks about and the "tools" he designed resemble tools as much as a brick resembles a Michelangelo sculpture. As for learning to think in a way that produces consistent results, this seems to be one of mankind's greatest challenges.

In a systemic attempt to provide a unified approach to economics, finance, and management, the Decalogue leads us inevitably to create the foundation for a new epistemology of wealth creation, something that through rigorous investigation separates knowledge (what we know and it is part of who we are) from superstition (what we believe is true and for which we have no basis of explanation, or only a shaky one).

In a world where the only cure for the folly of the financial markets seems to be stricter rules on the trading of derivatives and salary caps and where financial institutions still believe they can play dice because the government will eventually bail them out, we need to dig deeper on what causes the state we are in and find a way to come out of it. As ambitious as it may sound, that is the ultimate purpose and scope of the Decalogue (Figure 8.1).

8.2 Optimizing the System

The last 15 years have seen an exponential growth in the planetary interconnections of which we are all part. The artificial walls built by our inability to conceive and experience higher levels of interdependency are crumbling. It is time for the science of management to play its part in helping our lives to be meaningful in the work place.

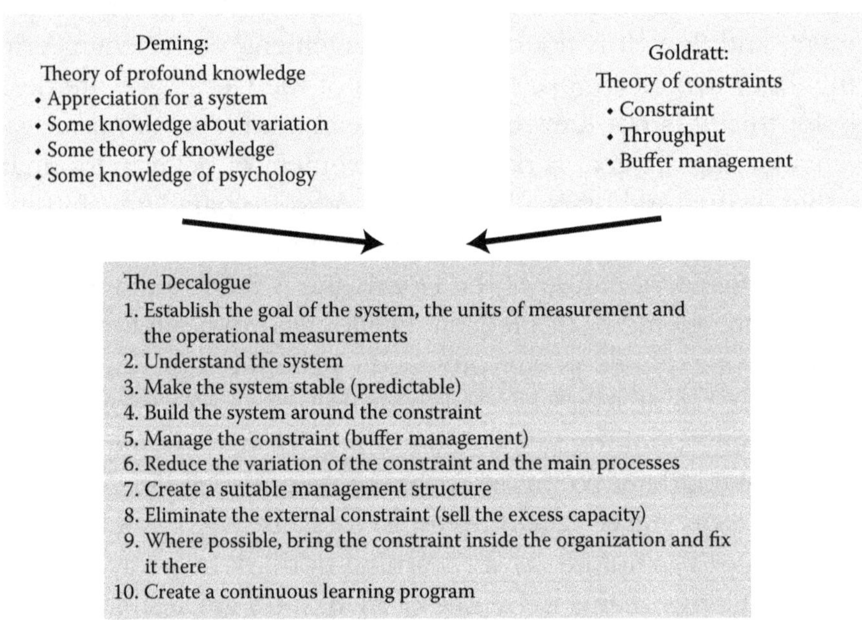

Figure 8.1 Elements of Deming-and-Goldratt's work unified in the 10 steps of the Decalogue methodology.

Philosophically and scientifically, the Decalogue methodology attempts to harness interconnections and create more meaningful practice. It does so by shifting management from a mechanistic, Newtonian worldview in which the results of the whole organization equal the sum of its individual, separate, and hierarchical parts, toward a systemic and interdependent network. It achieves this in practical terms by the following:

- Building interdependent processes managed through the control of *variation*
- Subordinating these interdependencies to a strategically chosen element of the system called *constraint*
- Designing the organization as a *network of interdependent projects* with a goal

The last 15 years of relentless practical application of these principles have yielded a new methodological approach in its own right. This has emerged from the interdependencies of practicing together the fundamental tenets of managing variation and constraints in a systemic way. Practicing the application of the 10 steps has highlighted their inner strengths and weaknesses, prompted new and deeper questions, and provided an opportunity for creating a much more integrated, cohesive, and defined algorithm. What emerged out of this process in over a decade is a powerful management philosophy that provides a cogent and consistent method for the transformation of the present style of management into one of optimization. *Optimization* means that every component of the system is optimized (and where appropriate, this means suboptimized in the traditional sense) to achieve the goal of the system.

8.3 What to Change, What to Change to, and How to Make the Change Happen

The three fundamental questions in TOC are what to change, what to change to, and how to make the change happen.

In the previous chapters, we have clearly identified "what to change" as the underperformance created by the prevailing style of management based on traditional hierarchical silos. We have also identified "what to change *to*" as the "chocked tube" model based on managing variation and constraints and the network of projects as the operational solution.

At this point, in order to answer the question "how to make the change happen" all we need is a protocol, that is, a detailed roadmap for the deployment of that solution. In the language of TOC, we need a prerequisite tree. This roadmap is contained in the 10 steps of the Decalogue and is, among other things, a very practical and useful guideline for the building of a systemic organization.

8.4 The Decalogue in Three Fundamental Phases

The 10 steps of the Decalogue are highly operational and in order for them to be fully understood it is useful to consider them within three fundamental phases.

Phase one: The system

The Decalogue was developed, scientifically and otherwise, within the framework of system management; a system is defined as a network of interdependent components that work together for a clearly stated goal. Accordingly, at the onset of any Decalogue effort it is mandatory to define as follows:

- The goal of the system we intend to operate (well beyond the boundaries of the legal entity that defines "the company")
- The units with which we measure the achievement of the goal (e.g., cash per year)
- The method we intend to use (the "how") to measure what we want to achieve
- The interdependencies (the links and linkages of the network) that allow the system to achieve the stated goal

Phase two: Managing the system through variation and constraints

The system must be managed. Perhaps the most relevant part of the monumental contribution that Dr. Deming gave to management, allowing this human endeavor to enter the realm of science, was his realization that any system in order to be managed must be understood in terms of the variation of its components, the processes that make up the system. Dr. Deming understood that, at the most fundamental level, the interaction of people, methods, materials, and the environment in which these elements exist, generates variation. By leveraging the rigorousness of the statistical work of Dr. Walter Shewhart, Dr. Deming created a unique and unrivalled management philosophy that provides, if properly understood and metabolized, the guiding elements for a transformation. Allow us to emphasize again that Dr. Deming's work, named by him at the end of his long life as the Theory of Profound Knowledge, is for the science of management what post-Newtonian physics is for our understanding of the physical world: a paradigm shift. Understanding and managing variation in the system is foundational for the successful adoption of the Decalogue and for the making of that shift.

About 30 years ago, Dr. Eli Goldratt started to point out, with as much lucidity as vehemence, an obvious but neglected fact: any system, organizational or otherwise, is intrinsically limited in what it can achieve by one or very few limiting factors. If we choose, strategically rather than opportunistically, these limiting factors ("the constraints") and take full advantage of (exploit) them by organizing (subordinating) the whole company to that purpose, we can maximize what the system can realistically achieve toward its stated goal. Dr. Goldratt has amply illustrated his view on how organizations can achieve their goals following his approach in the vast body of knowledge he has developed named TOC. Dr. Goldratt has developed a wealth of very powerful applications for the proper utilization of the finite element(s), the constraint(s) that limit organizations in achieving their goal and has trained a whole host of instructors capable of teaching them.

In the late 1980s, Dr. Goldratt started to develop what he called the "TP tools" and described them in a novel called *It's Not Luck* dedicated to illustrating how it is possible to systematically overcome "constraints" that are not physical.

TOC has been nothing short of a revolution. Needless to say, the applications of TOC to managing the constraint are critical for an effective implementation of the Decalogue. Just like the most famous offspring of the theory of variation, SPC, these applications need to be fully understood in order not to become an expedience or merely "techniques."

Phase three: Grow the system in a sustainable way

A system must grow in its ability to achieve its goal. This growth can only happen in a sustainable way if the people that work in the system and the people that work *on* the system are capable of learning about methods, products and markets in a continuous, systematic, orderly manner and are willing and able to apply their learning. The organizational design that supports this continuous growth through learning is probably the most complex and least understood aspect of the systemic approach to the management of organizations.

To summarize: The basic elements for sustaining the long-term development and success of an organizational system are the following:

- Clarity on the goal and the network of interdependencies that enable its achievement
- Understanding and management of the variation associated with each process and their interactions as well as identification, exploitation, and subordination of the system to a set of limiting factors, the constraint(s) of the system
- Establishing an organizational mechanism that ensures appropriate capitalization on people's continuous learning about methods, markets, and products.

8.5 The 10 Steps of the Decalogue

In this section, we provide some commentary on the 10 steps of the Decalogue in further detail.

Step one: Establishing the goal of the system, the units of measurements and the operational measurement

Without a goal there is no system and without clarity on what to measure in the system and how to measure it, talking about a goal becomes lip service. GAAP (Generally Accepted Accounting Principles) accounting utilized to support managerial decisions is the enemy number one of productivity. EBIT (Earnings Before Interest and Tax), EBITDA (Earnings Before Interest, Taxes, Depreciation and Amortization), EPS (Earnings Per Share) and any form of GAAP-derived measurements totally miss the point of what a company should strive for. If the goal of a company is connected in any way with making money, then all we need to know is the following:

- What comes in (sales)
- What goes out to purchase materials and services that go into the products we sell (*TVC*, totally variable costs)
- What we need to make the system function (fixed costs + investments), operating expenses (*OE*)
- The inventory (*I*) we need to keep in the system to ensure that we always have enough "material" to produce and ship

These basic variables are connected in the following way:

Sales minus TVC = throughput (*T*); *T* minus *OE* = net profit; *T* minus *OE* minus *I* = cash profit, the physical money we see in the bank account before tax. Indeed, from year to year, *I* becomes Δ *Delta I*.

Throughput is overwhelmingly more important than inventory and *OE* because it can potentially grow far more than *OE* and *I* can be reduced only so much. Sadly, the whole GAAP effort revolves around "understanding *OE*" with devastating effects on decision making. Let us say this clearly: the industrial world is where real wealth can be created and industry needs to learn how to measure its performances based on *T*, *I*, and *OE*. Banks, the stock market, financial institutions, and the guild of accountants *must* understand that their job is to support true wealth creation by fostering in industry the pursuit of these measurements instead of thrusting upon them anything different.

Nearly 30 years ago, Dr. Goldratt explained these simple concepts and all the financial and industrial catastrophes we have experienced in the last 25 years are, directly or indirectly, the result of not having understood them. Throughput is the most powerful way to link the industrial and the financial worlds and the accounting that goes with it provides a whole new and meaningful scope to the endeavors of the accounting profession.

It is also critically important to measure how the system performs in its effort to deliver to customers. We can have a pretty good handle on it by measuring the on-time delivery and assigning a dollar value to any delay, we call it *T$D* (throughput $ day). The longer the delay, the higher the *T$D*. Similarly, we want to know how much cash we keep trapped in inventory to achieve an optimal, ideally zero, *T$D*, and we call it *I$D* (inventory $ day).

Step two: Understand the system (draw the interdependencies)

If we do not know who does what, what the inputs and outputs are, and how everybody's work is connected, then we are not managing. As Dr. Deming used to say, "Business schools teach you how to raid a company not how to manage it." It is disconcerting to discover how little we know about our system until we have a clear picture of how our interdependencies are laid out and how neglected by top management this issue is. Through mapping out all the main processes of an organization, step two provides the foundational elements of understanding that will enable the building of a truly effective quality system.

Step three: Make the system stable (understand variation and its impact on the network)

Dr. Deming used to say that if he had to reduce his message to management to just a few words, he would say that all their job is about is *reducing variation*. Dr. Deming's unrivalled contribution to the science of management comes from having understood the importance (and all the implications) of a body of statistical studies developed by Dr. Walter Shewhart of Bell laboratories in the 1920s. Shewhart found out that any process is affected by variation, which is enemy number one of quality and reliability; however, this variation can be attributed to either common causes or special ones. Distinguishing "noise" from "signal" was then critical to devising actions aimed at managing this variation. Dr. Shewhart developed an important part of the theory of variation known as SPC and a very useful mechanism for measuring variation called a control chart. His work was foundational for the improvement of productivity, first at Bell and then in a myriad of organizations countrywide, and certainly served as a springboard for the gigantic work of Dr. Deming.

Dr. Deming realized that in order to manage the variation associated with a process people make two kinds of errors:

- They attribute variation to a special cause when it is instead due to a common cause
- They attribute variation to a common cause when it is instead due to a special cause

The ramifications of these errors are endless and still today plague the way companies make decisions; SPC, the main body of knowledge out of which the TPK has evolved, is largely ignored at top management and corporate level. SPC is neither a purely mathematical tool nor is it a conventional financial management tool. The process behavior charts mirror the outcome of our managerial decisions, often taken with some kind of local optima in mind. The image they portray is not the reassuringly deterministic one that accountants and financial people are accustomed to; on the contrary, what these charts display are predictable or unpredictable ranges of oscillation. These oscillations very often reflect, mercilessly, the conflicting confusion that dictates our choices and the course of action that their interpretation calls for flies in the face of "conventional wisdom."

Understanding SPC, let alone using it properly and accurately, requires a paradigm shift in the way we look at data and make sense of them for business decisions. From a mathematical standpoint, process charts are based on an average dispersion statistics and the approach they use is the 3sigma one for the calculation of limits. What has always generated confusion about SPC charts is that although connected with probability theory they do not work *because of* it. The essence of the charts is in their predictive role and in the possibility they provide to build an epistemological approach to management based on prediction.

Control charts capture the most fundamental feature of the work of individuals and their interaction within an organization: the variation associated with processes. In building a systemic organization based on the Decalogue, SPC has a central role.

Step four: Build the system around the constraint (subordinate the organization to the constraint)

The new kind of organization that is based on managing variation and constraints is a network of interdependent processes with one common goal where we have achieved a good level of statistical predictability. It can be successfully managed, but the question is, how? Dr. Goldratt's main contribution to the theory of management has been to point out that any system is limited toward its goal by very few elements, the constraints. If we identify them we can manage them following the steps of focusing that he developed. The Decalogue, leveraging the intrinsic stability of a Deming-based system, suggests that the constraint can be "chosen" (one constraint) instead of being identified. In other words, we can always decide which constraint it is strategically more convenient to focus on and build the system accordingly. However, we can do it *only* because we have already built a system made of low variation processes; this is why we can safely design our company around a strategically chosen constraint. Instead of cycling the five focusing steps of TOC—(1) identify the constraint, (2) exploit the constraint, (3) subordinate to the constraint, (4) elevate the constraint, and, if the constraint has moved, (5) go back to step (1)—we can make the system grow by appropriately choosing the constraint and sizing the capacity of all the feeding/subordinating processes coherently. Again, this is possible only because the variation in the system is low.

Step five: Manage the constraint (protect and control the system through buffer management)

The performance of this system is ensured by its predictability but we need a mechanism to protect it and control it. This is provided by the buffer and by its management. What is a buffer? Individual processes exhibit variation; two or more together do too, this is called covariance. The effect of covariance is a cumulated variation that can result in any sort of combination of these variances. Regardless of how little individual variation the processes of the system have, we do need a mechanism to protect the most critical part of our organization, the constraint. A buffer is a quantity of "time" that we position in front of the constraint to protect it from the cumulative variation that the system generates. Simply, in synchronizing the processes that deliver the output of our organization, we ensure that what has to be worked on by the constraint gets in front of it "one buffer time ahead."

Is there a precise and unequivocal recipe to size the buffer appropriately? No, and it doesn't matter. If you understand control charts, the sizing is fairly obvious. The real importance of the buffer is not in the protection from disruption, certainly not an irrelevant issue, but in the possibility to exert control over the functioning of the whole organization: if our processes are all predictable, then we can truly use the buffer as a mechanism to gather succinct, effective, and comprehensive information about the "state of synchronization" of our company. Indeed, if we monitor the buffer with the use of a control chart, then the control becomes real insight, the goal of management.

The management of the buffer entails a Copernican shift in the way we mean control because it prompts decision makers to rely on understanding and knowledge rather than hunches; it forces them into leveraging arguably esoteric statistical studies instead of salt of the earth accounting; it constrains the unbridled vibrancy of managers into the straightjacket of rational thinking. As a result of those considerations, over the decades, many of the repeated attempts to introduce buffer management to top management have failed.

Step six: Reduce variation of the constraint(s) and the main processes

Step six is obviously connected to step five but less obviously to step seven. Clearly, we do understand the impact that variation has on our system and the need to reduce it but, when push comes to shove, we are not prepared to continue to work on variability reduction. Why not? The answer is in our ability and desire to understand the purpose of system management.

The culturally disappointing translation of Deming's Philosophy into a myriad of "Kaizen-like" management techniques has transformed it from a vision of the world based on innovation and wealth creation into an efficiency game. If, on top of this, we continue to view our company in "functional" terms, then reducing variation simply means reducing costs. Of course, no function would ever easily surrender to that because it would imply "cutting the budget," hence any serious attempt to reduce variation is nipped in the bud. A relentless effort toward continuous reduction in variation can only stem from a *systemic* vision of our company and the understanding that only this reduction would provide the insight needed for triggering real jumps in performance. The way to link a relentless, focused and companywide variation reduction crusade to financial performance is through the adoption of a suitable organizational structure.

Step seven: Create a suitable management/organizational structure

At the time of publication of the book *Deming-and-Goldratt: The Decalogue* (1999), step seven was generic.

It was clear that, without a suitable structure, the realistic possibility to sell all the capacity of the constraint would be hindered by local optima considerations. In other words, the design of a suitable structure was a prerequisite for enabling the true expansion of the system. Today, it is much clearer that the natural way to see the work of an organization is in terms of process and projects. In the years that followed, further research and application led to the development of the *network of projects* organizational design as a means to improve and optimize the performance of the whole system.

Step eight: Eliminate the external constraint (sell all the capacity the system has available)

When we design a system that caters for a high degree of process predictability and synchronization, where control and protection are ensured by buffers, and where all the policies, behavioral, and measurement "constraints" are dealt with by an appropriate organizational structure, we do so to maximize sales. The most important part of the chain is the customer and any company should always be designed to ever improve its ability to satisfy its customers' verbalized and hidden needs. The Decalogue, if understood and embedded in the appropriate structure, should very quickly reveal capacity that is not currently being sold.

Another way of looking at this issue is the following. Let's say that, day one, the rate of sales of a company starting its Decalogue journey was such that some shipments were missed and constant firefighting would create friction between production and sales. The Decalogue would call for a disciplined process mapping effort aimed at understanding process variability, the choice, and management of a suitable constraint as well as the devising of a coherent measurement system. Moreover, in order for this level of synchronization not to be hindered by the local optima inclination that functional organizations invariably undergo, we would design an appropriate, coherent organizational structure. Almost invariably, the constraint would shift outside, thus becoming an external constraint: our capacity, what we can realistically design, manufacture, and ship, would become greater than what we were currently capable of selling. At this point, it would be blatantly obvious that our real understanding of the market is woefully limited and we, in truth, do not know how to sell. Dr. Goldratt developed a powerful and detailed TP approach to managing the external constraint. This is a very critical point in the pattern of a successful Decalogue implementation. Why? When left to behave in a traditional local optima way, salespeople can singlehandedly jeopardize any systemic endeavor unless we integrate them organically into the way the company operates.

Step nine: Where possible, bring the constraint inside the organization and fix it there

The Decalogue approach to management is based on process stability; indeed the most critically important part of the system is the constraint. Hence, we want to ensure maximum predictability especially on the constraint. Clearly, when the constraint is external such predictability is more difficult to achieve. This is the reason why, whenever possible, we want to manage an internal constraint. Moreover, this is also the easiest way to make the system grow without stirring chaotic company dynamics.

The need for constraint reliability is so strong that even when organizations have a virtually unlimited internal capacity, like supermarkets, we should always elect to appoint an internal constraint and subordinate the whole organization to it. The growth of the system would then happen through a systematic, orderly and relentless exploitation of the capacity of the constraint. When such a capacity is not sufficient to meet market demand, we would first increase the appropriate non-constraint areas to make them capable of subordinating to the constraint and only then would we elevate (increase the capacity of) the constraint. Once again: the name of the game is process predictability.

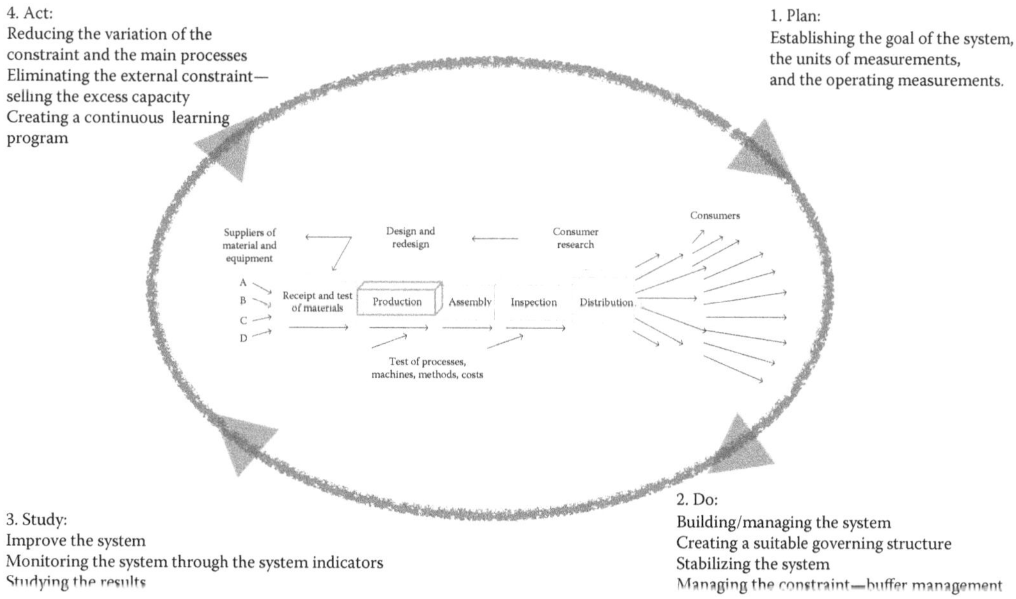

Figure 8.2 Steps of the Decalogue mapped onto the PDSA cycle.

Step ten: Create a continuous learning program

The possibility for a Decalogue-based management system to produce results over time rests on the ability that the organization has to continually learn what is needed to constantly improve its performances.

Learning does not happen in a vacuum and cannot be based solely on individual desire. Learning and personal development must become part of the way the organization functions and the change associated with it must become a way of life for the company. Learning cannot be "installed" neither can it be forced on people. It must be a personal choice but also an integral part of the way the company has structured itself to conduct business. Learning and the self-development that comes with it must be promoted companywide and from the top management but must come from a designated and empowered source. Over the years, we have come to call it a center for learning.

The Decalogue is founded on the principle of continuous improvement. Not only does it employ the "Plan, Do, Study, Act" continuous improvement cycle designed by Dr. Deming, the entire 10 steps embody that cycle pattern in the way they are carried out. This can be seen clearly when we map out the 10 steps of the Decalogue onto the PDSA cycle itself (Figure 8.2).

Chapter 9

Why Do We Get Stuck on the Path of Transformation? The Inherent Conflict of Change

> Everyone doing his best is not the answer. It is first necessary that people know what to do. Drastic changes are required. The first step in the transformation is to learn how to change...Long-term commitment to new learning and new philosophy is required of any management that seeks transformation. The timid and the fainthearted, and people that expect quick results, are doomed to disappointment.
>
> **W. Edwards Deming**

9.1 Change and Consciousness

We live in a world that is changing rapidly, where individuals and organizations are entangled in an ever-increasing web of relationships, in an evolving network of conversations that keep us all more and more connected. The complexity generated by these interdependencies produces emergent properties that cannot be understood only in terms of the constituents, hence making futile any attempt to govern them by breaking them into pieces. This is the lethal fallacy of the hierarchical structure, the functional organization, and the cost accounting-based methods that go with it. If we want to replace the conventional hierarchy with a model that addresses all the shortfalls induced by it we need look no further than a systemic model guided by the principles and methods set forth by TOC and the TPK. A cohesive and coherent way of doing it is provided by the Decalogue and in the previous chapter we examined the 10 steps to create the transformation.

Even a much less powerful organizational design than the one originated by combining high reliability and effective synchronization would yield a more viable organizational structure than the functional structure. Every day, every industry experiences the drawbacks of a functional structure. The associations of quality professionals and the registration bodies, while promoting a loosely defined system approach to management, always refer to the hierarchical structure and seek better and better gimmicks to minimize the damages.

We do not lack the knowledge needed to build and operate a truly systemic company guided by the idea of throughput; what we lack is the ability to *decide* to use that knowledge. The ability to choose what is "right" goes hand in hand with the ability to *develop and reveal our consciousness*. Any time we have to make a decision, thus triggering a change, we have the need to "name" that change in a way that we perceive it as "good" for us. We have to perceive the value of our choice at every level of our consciousness: intellectual, emotional, and pragmatic.

The successful adoption of a systemic approach to management must address in an unprecedented way the people's consciousness, the deeply rooted images that they build about themselves, their role in the company, and their perceived status in the world.

As sad as it may seem, people accept the nonsense and frustration they have to cope with in a functional company because they can relate to it. And so they accept the progressive separation between who they think they are and what they do. Such an accepted separation deteriorates into an even worse disconnection: the one between what we learn and what we accept to change as a result of that learning. It is time for that to stop.

9.2 The Constancy of Change

Our pancreas replaces most of its cells every 24 hours; the cells of our stomach lining are reproduced every 72 hours; our white blood cells are renewed in 10 days; 98% of the protein in our brain is turned over in less than 30 days; our skin replaces its cells at a rate of 100,000 cells/min.

Change is a defining feature of life. Simply put: change happens. We can look at change in many different ways but it all boils down to how we deal with it. We can ignore it. We can decide that change is something that will never truly affect us, our families, our business, or our country. We can decide, "Our way of life is not in discussion." This would be understandable (although not desirable) because change does require a considerable cognitive effort. Dr. Deming used to say, "Learning is not compulsory, but neither is survival."

Or, we can accept it for what it is, live with change or even thrive with change. If we decide to do so, we can read the myriad of books that every year are published on this issue and glean an understanding of all the angles that change can be tackled from. Undeniably, this is a multifaceted issue and different realms of knowledge are involved in providing this understanding. The conflict "Change versus Don't Change" is probably one of the most profound issues of our times.

9.3 Looking at the Conflict Cloud of Change

Ignorance is a condition, stupidity is a choice.

Wiley Miller

Why do we get stuck on a transformative path? Why it is so difficult to change, even when we know how to make it happen? Why is it so difficult to embrace newly learned and valid concepts and apply them? We can analyze this situation of blockage or impasse with the Conflict Cloud, the way we framed questions about complexity and hierarchy in previous chapters. Here, we will look at a very fundamental conflict cloud connected with change (Figure 9.1).

We can read both sides of the conflict as it is depicted in the diagram in words:

If the goal is to work toward a desirable future (A) then we must achieve security (B) and therefore we do not want to change (D).

On the other hand, if the goal is to work toward a desirable future (A) then we must pursue satisfaction (C) and therefore we want to change (D′).

In TOC, we refer to conflicts as "inherent" when they typify a situation that is foundational for the issues it creates. These are "paradigmatic" conflicts that can be experienced in a myriad of ways but they can all be boiled down to one, essential conflict.

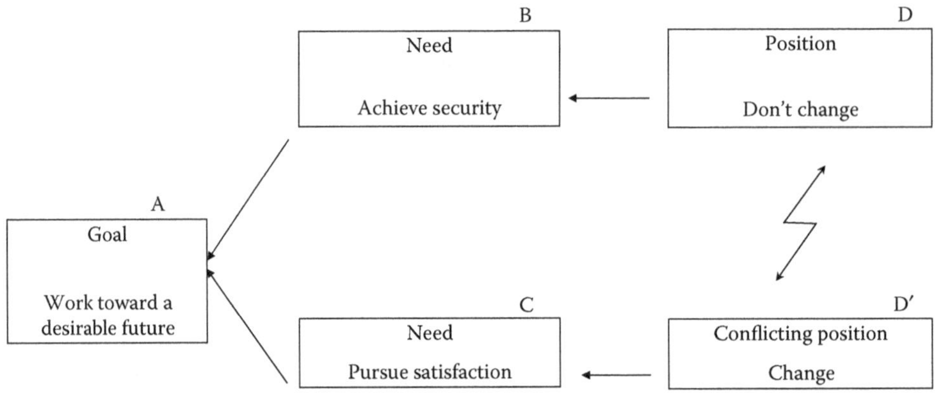

Figure 9.1 Inherent change conflict cloud.

When we come to the very essence of change, a plausible way to verbalize the paradigmatic conflict of change is the following:

need B = security
need C = satisfaction
goal A = fulfilling life

Indeed, "security" and "satisfaction" are very generic verbalizations of the two deepest needs that shape our existence; security is connected with fear, control, stability, restraint and, in general, deals with that side of the human experience that philosophers would call "immanence," that is, what pervades our existence here and now and that we call "reality." Satisfaction, instead, is linked to desire, vision, tension toward what exists not only in the reality we experience in the present but in what we perceive to be possible in the future. It is grounded in the hardwired synaptic circuitry that propels us all toward what "transcends" us.

The drive humans have for transcendence is an innate desire that we all share to go beyond our current state and see ourselves projected into a different and greater dimension of existence. This is the reflection of an often untapped level of consciousness. This is what many call "soul," or that part of us that aims high, that is projected above, that calls for new challenges and yearns for a higher level of spirituality and meaningfulness. In the very same way in which we need to exert "bodily" control over our environment we also need to project a "soulful" vision of ourselves into the future.

Security and satisfaction, or whatever verbalization we use in any specific case, are the main emotional drivers that, sometimes rationally and more often subliminally, inform the choices we make. Security and satisfaction, alternatively we could say restraint and expansion, are "emotional attributes" that take their shape as a result of our experience and sit very deeply and inaccessibly in our limbic system. If we could access them and see them more clearly it would be much easier to accommodate the real needs that drive people's behavior. We are getting there. When functional MRI technology produces all the wonders it seems capable of, the science of the brain will have gone up a notch and we will know better how to help people with all kinds of mental disturbances. When applications of this newly gained, science-fueled knowledge are developed to aid organizational development, we will have a much stronger means to support the true development of human capital.

9.4 Change and Leadership

The conflict of Change versus Do Not Change as captured in the conflict cloud is of huge importance because every single one of us experiences it in both minor and major matters every day and every single one of us experiences it differently depending on the deep seated needs (the B and C of the conflict cloud) that we have. It goes almost without saying that the "Change versus Do Not Change" conflict of a leader is the lens through which they will read every situation. Why is it so important to acknowledge this? Because wittingly or not, it will inform their decisions. Leaders, more than anyone else, need to be able to separate legitimate mental models based on some sort of facts from *personal bias*. One way of doing that is to keep a constant check on the "state" of their personal change conflict.

Any conflict rests on unchallenged mental models, or what in TOC are called "assumptions." Dr. Goldratt gave the name "injection" to any statement that

1. Logically invalidates one or more assumptions and
2. Ideally addresses both needs in the conflict cloud (or only one while being non-detrimental for the other). We can verbalize some completely generic assumptions in this conflict of change as follows:
   B–D: What we have done has worked so far; the majority of our people would find it difficult to absorb the changes; we cannot assess the risk associated with the proposed changes, etc.
   C–D′: We know we can do fundamentally better; competition will not sit idle while we rest on our laurels; there are amazing opportunities out there which we are not tapping into; we cannot let complacency take over.

D–D′: Change is a choice; change is avoidable; there is no silver bullet for change; there is no way to predict the results of a change; if we change we are not ourselves anymore, etc.

When we read out the full conflict from left to right, it comes out like this:

If the goal is to work toward a desirable future, and the future is threatened by loss of control, and loss of control threatens security, then we must achieve security. And if we must achieve security, and what we have done has worked so far; the majority of our people would find it difficult to absorb the changes; we cannot assess the risk associated with the proposed changes, etc., then we *do not* want to change.

On the other hand, if the goal is to work toward a desirable future, and the drive toward the future is triggered by satisfaction, then we must pursue satisfaction. And if we must pursue satisfaction, and we know we can do fundamentally better; competition will not sit idle while we rest on our laurels; there are amazing opportunities out there which we are not tapping into; we cannot let complacency take over, then we *want* to change.

Why are we stuck in the conflict of change versus not change? Because we make a series of assumptions (D–D′) including:

Change is a choice; change is avoidable; there is no silver bullet for change; there is no way to predict the results of a change; if we change we are not ourselves anymore, a comfort zone exists and can't be modified in a planned way, the future is something that happens to us, etc. (Figure 9.2).

9.5 Core Conflicts

Change versus Do not Change belongs to the restricted circle of the "core conflicts," just like the complexity cloud we looked at in Chapter 1 (we use the term "conflict" and "cloud" interchangeably and try to stay as much as possible within the conventional terminology of TOC). Indeed, any organization, and any individual for that matter, has their own "core conflict" but, almost invariably the two needs underpinning that core conflict fall into the category of security and satisfaction. As such, it is relatively straightforward to come to terms with the core cloud of any organization (and individual). We shall elaborate further on the application of the cloud and the TP tools as invaluable aids to any decision maker seeking a genuine systemic transformation. For now, it is important to understand the organizational implications of looking at an organization through this lens of the change conflict.

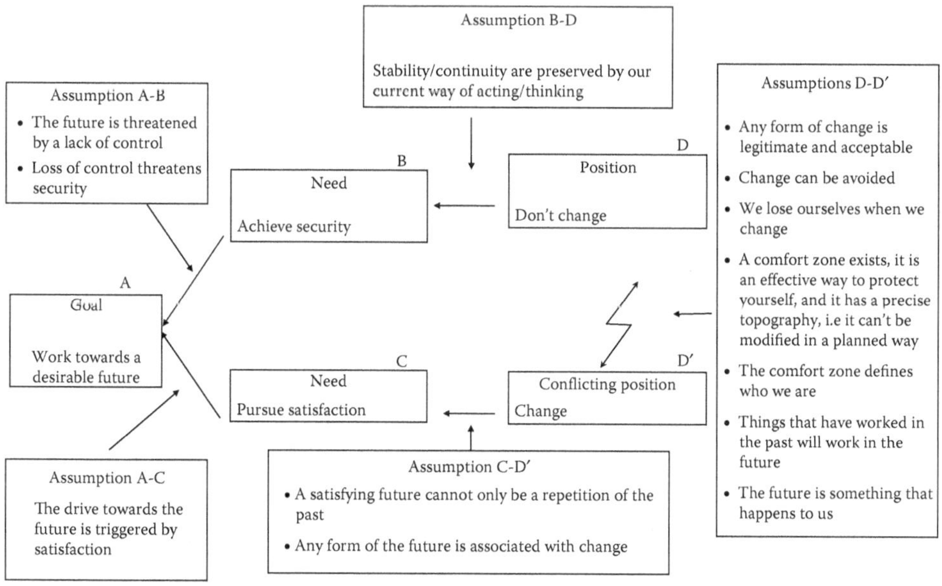

Figure 9.2 Change conflict cloud with assumptions.

In the following chapter, we will look at the TP tools that can help us with the various phases of change. Those phases of change, and the supporting TP tools that go with them, mirror the rational effort that we need to make in order to connect our *intuition* (for what needs to be changed) with our *understanding* of the new future scenario (what are we changing toward) and the awareness of the right *knowledge* we need to have in order to make that future reality happen (how to cause the change).

Whatever efforts we make to understand and cope with change, the most fundamental lesson to take away is that it is never really a choice about changing or not changing. Change is not something we can accept or refuse. Change is inevitable, and the sooner we learn to thrive with it the sooner we can learn to manage within complexity.

9.6 Levels of Resistance to Change

Any time leaders try to introduce change into an organization, they will be faced with the fear and conflicts of those involved in the changes. If these reactions are not managed, they will remain real and present and disrupt the making of changes. This is one of the most delicate, intangible and crucial aspects of leading and managing change. It is a cognitive challenge and it requires attention at the level of people's thoughts and emotions.

Thanks to the depth of thinking and research carried out in a semi-empirical way, Dr. Goldratt was able to identify *six levels of resistance to change*. These levels depend on the extent of awareness and participation of the people involved. When we are able to identify which level of resistance corresponds with people's reactions, we are well on the way to resolving the problem. Moreover, for every level of resistance there is a TP tool that will address and resolve that resistance in a collaborative way (see next chapter).

Level 1: Disagreement about the problem

This is the deepest level of resistance. At this stage there is no common agreement about what the problem is that is to be addressed through some form of change. To tackle this level, we need to identify the root cause of the majority of Undesirable Effects or UDEs that are being experienced. This can be done very effectively and quite fast through the *core conflict cloud*, starting from listing all the UDEs, or symptoms of the problem that needs to be addressed. No matter how complex the problem, through this analysis there is consensus about what the problem is and the need to find a solution.

Level 2: Disagreement about the direction of the solution

Once there is agreement about what it is that needs to be changed, there has to be agreement about how the change will be implemented. The direction of the solution is found by identifying "*injections*" to the core conflict cloud. This is done by invalidating the assumptions (mental models) that keep people stuck in the core conflict. It allows a shift toward a stage of agreement where it is no longer "you against me because of the problem" but "you and me against the problem."

Level 3: Lack of agreement that the solution will bring the expected benefits

This level reflects a disconnect between the understanding of change required and the logic that will bring the results. A fully fleshed out solution needs to be mapped out. For this reason, we build a *future reality tree*. This tool gathers together all the "injections" with a supporting logic to prove that the proposed changes will bring results. It will also lead to the identification of further injections that complete the solution. The result is a clear map that becomes a living document shared among those involved in the change process.

Level 4: Fear of negative consequences generated by the solution

People will be sensitive to possible negative implications they perceive for themselves through the implementation of the solution. It is important to identify negative implications early in order to protect the proposed changes from unexpected difficulties, to get buy-in from people who may tend to be nay-sayers and to demonstrate leadership through an ability to listen to relevant objections and incorporate that feedback. The TP tool for this is called *negative branch reservation*.

Level 5: Too many obstacles along the road that leads to change

At this level, the solution is much closer to being accepted as reality and the objections are at a very concrete level of obstacles standing in the way, often in the shape of a lack of resources. The people who see the obstacles are very often the very people with the knowledge to overcome them. It is vital to be able to address these obstacles and overcome them. The step to overcome an obstacle is called an *intermediate objective* (IO) and all the necessary IOs can be mapped out on a *prerequisite tree* in a logical order of what needs to be done before another IO can be achieved.

Level 6: Not knowing what to do

When level 5 has been overcome we are in a situation where all that needs to happen is for actions to be taken to achieve the IOs. Even so, at this stage people may freeze. This may be simply because there is not enough clarity on tasks or because not all the objections have been raised. This is a critical phase in terms of leadership and can be greatly facilitated by giving and sharing clear instructions through the use of the *transition tree*.

	Level of resistance to change	Thinking process tool
1	Disagreement about the problem	1. Conflict cloud
2	Disagreement about the direction of the solution	2. Injections
3	Lack of faith in the completeness of the solution	3. Future reality tree (FRT)
4	Fear of negative consequences generated by the solution	4. Negative implications branch
5	Too many obstacles along the path that leads to change	5. Prerequisite tree (PRT)
6	Reservations about our ability/willingness to implement the solution (and about the ability/willingness of others)	6. Transition tree (TRT)

Figure 9.3 Levels of resistance to change and the TP tools to address them.

The TP tools promote and reinforce the kind of systemic intelligence needed to create and manage an organization as a whole system. They provide a complete and holistic approach in a continuum from high-level strategy through to detailed step-by-step implementation. They are both the means and the method to create, communicate, and share any process of change. They reinforce and cultivate the true involvement of people and are therefore crucial for improving flow in communication which, in turn, enables better flow in every sense (Figure 9.3).

Chapter 10

The New Intelligence: Intelligent Emotions and How to Foster a New Systemic Intelligence

10.1 Systemic Thinking to Inform Decisions and Minimize Risk

In our extraordinarily complex world, interdependencies and interconnections multiply at an ever-increasing speed. The cause–effect relationships that govern the world as we experience it form a super intricate "network of networks" and we have a very limited understanding of the underlying properties of these networks and the laws that govern them. For this reason, never like today, with interdependencies multiplying, do leaders and managers need to be able to literally "connect the dots." They need to be able to see and build the right cause-and-effect connections that shape and influence reality. The systemic thinking approach to analysis and action presented in this book will allow them to make *informed decisions* and, as a by-product of the right kind of thinking, *minimize risk*. It will also allow them to coexist and even thrive with the continuous level of change, adaptation, and continuous innovation required to address complexity.

In this chapter, we look at a set of TP tools from TOC that allow leaders and managers to foster the systemic intelligence they need. These tools will also allow them to harness their intuition and deepen their understanding and knowledge. This, in turn, will influence their emotions so that they can become "intelligent emotions." When we reach a level of intelligent emotions, those emotions become a positive force to be employed toward the goal as opposed to a minefield of potential discord and stress.

10.2 The Cognitive Constraint: Relearning to Think

It will not occur to many people that they need to learn how to think. Surely it is something that comes naturally as we do it all the time, and, apart from some exercises in basic mathematical logic, it is certainly not among the skills we learn at school. However, our brains grasp at straws any time we have to come to terms with the implications of the nonlinear laws that govern networks. Today, very sophisticated computational models help us to penetrate complexities by connecting simple structures. This allows us an increasingly profound understanding of complexity. We already have the knowledge and the tools to do so much better than we do, but we continue to lag behind that potential.

If we want to close the gap between the knowledge available and what we are willing to use, we have to tap into a different use of the mind. We have to learn how to see change as not simply something to be feared, but a natural, intrinsic part of our life. When we learn how to use our intuition and intellect to implement consistent action, change is no longer a threat and a hazard; it is a continuous source of new opportunities.

That possibility is connected with a new and more powerful ability to use the three faculties of the intellect connected with intuition (birth of an idea), understanding (analysis/development), and knowledge (application/execution). Tapping into these largely unexploited faculties increases our ability to leverage our intelligence and our ability to learn and act consistently with this learning.

The issues then become as follows:

- How can I fortify my intellect?
- How can I develop a better ability to generate intuition, understanding and knowledge?
- Is there a practical way of doing it?

Gaining a better, more disciplined handle on the way we think, speak, and act requires a non-negligible effort. This effort can be fuelled and sustained only if the achievement of this new mastery addresses simultaneously the two main needs that both individuals and organizations have: control and vision.

Organizational transformation requires better thinking and learning. This transformation, of course, cannot be disjointed from the transformation of the individuals that belong to it, but organizations have their own set of paradigms that are the offspring of the way interdependencies are created and managed. In other words, organizations have a life of their own that is generated by the somewhat unknowable combination of the individuals inside them and the way they interact with each other, what we may refer to as the emergent properties that arise from this combination. It is this complexity that calls for a new, stronger, and deeper insight.

10.3 The Cognitive Shift

It all comes down to a matter of choice. We can *choose* to replace the outdated hierarchical/functional structure with a much more organic design, based on managing variation and constraints, that reflects the intrinsically project-like nature of the work of organizations. This change is not cosmetic, it is transformational and it is rooted in a paradigm of cooperation, togetherness and win–win. The new covenant that everyone in the organization (as well as the value chain in which the organization is embedded) must embrace requires a much higher ability to think, communicate, and act; it requires a new "wiring" in the way we measure, manage, and sustainably improve our efforts toward our goals.

The real challenge in bringing about a transformation based on a network of projects lies in the emotional and cognitive shift that needs to occur in the way people learn and use their knowledge as well as how they see themselves, develop, and interact in the workplace. The TP tools from TOC were developed by Dr. Goldratt to aid that shift. Far from being a mere technique, we have found these tools to be a critical element in transforming organizations into "thinking systems" because they foster in people the ability to see interdependencies, resolve conflicts, be empowered, work cooperatively, communicate effectively and, at a higher level, perceive (and act coherently upon) the oneness of the organization with its business and physical environment.

To shift successfully from a traditional hierarchy toward a network of projects, we need to start from the very foundation of how we think, speak, and act. What we need to learn is how to seamlessly connect our uniquely human abilities to the following:

1. Systematically develop solutions to seemingly insurmountable conflicts (intuition)
2. Understand the full spectrum of these solutions (analysis/understanding)
3. Develop and implement a cohesive and cogent action plan to implement the solutions (knowledge)
4. Sustain the development of the "intelligent emotions" needed to harness complexity

10.4 The Tools We All Need

Dr. Goldratt developed the Thinking Process Tools to address constraints that are *not physical.* Nonphysical constraints are all of those limitations that are created by our mind as a result of past experiences. Some of these constraints are very helpful and make our life possible; restraints can play an extremely important part in our lives just as much as the possibility to lift them when appropriate.

These limitations do not belong only to individuals but also to organizations; the field of forces created by the structure of the organization, the paradigms of its founders and directors, and the sociopolitical environment at large do shape these constraints. The issue, for individuals and organizations alike, is not so much the existence of these constraints but rather the acknowledgement that they are such; unfortunately, far too often, these mind-created constraints become the "reality" of the organization. The life cycle of companies, and even entire industries, can be measured in terms of their appreciation (or lack of it) for these deeply rooted images of the world that over the years we have come to call "cognitive constraints."

The TP tools fortify in people the ability to reason cause-and-effect, a daunting task as our minds simply do not work that way. Recent studies in neuroscience are a clear indicator of that, demonstrating how much our thinking and decisions are dominated by the "lizard brain." In our daily lives, most of the time we "re-act" instead of acting and we very rarely understand the full spectrum of the consequences of our "re-actions." Simply put, human cognition is heavily constrained in understanding the network of interdependencies that our actions trigger. Dr. Deming used to warn about the consequences (cause-and-effect) of seemingly simple actions: "If we kick a dog in the street, we are responsible for the attack that, probably out of fear, the dog will take against the next passerby."

The TP tools sustain and focus the change process underpinned by the process of ongoing improvement at the heart of TOC during the three major phases of change: (1) What to change; (2) What to change to; (3) How to make the change happen.

Goldratt realized how important it was to provide a strong cognitive support to combat the difficulties associated with change that people inevitably experience. Used as a complete suite, the TP tools have proven to be an effective way to supervise and guide the change process. They are also an ideal companion for developing project plans.

Using precise verbalization and simple diagrams, the TP tools help people to visualize the complex, highly nonlinear network of cause–effect relationships that mark reality, as we perceive it. They allow us to map the "conversations" that make up our cognitive horizon, and these conversations are in themselves a kind of network. Without appropriate tools, we have no way of grasping the reality of this network, how it defines the semantic boundary of our universe, and how it shapes our actions. Learning to use these tools takes some practice, but within days they allow people to focus and accelerate their work together. Over time, they enhance and fortify the faculties of the intellect that are responsible for the conception of new ideas (intuition), their full development through analysis (understanding), and the operational deployment of the actions needed to carry out the implementation of the fully analyzed idea (knowledge).

By linking these faculties, the tools enable a higher level of control over the interdependencies among these faculties, and this can often lead to conceiving breakthroughs that did not previously seem possible. The tools also have an important role in reducing variation in our thought processes by focusing our mental efforts toward a goal. They can greatly reduce variation in the way people communicate in an organization by providing a common language. Moreover, they help to harness the powerful forces represented by the emotions involved in the change process, and reinforce and engender the empathy required for collaborative work. The learning required through a change process can be very destabilizing on an emotional level because it continuously pushes forward the limits of our cognition. This means creating a gap between what we know and what we feel we can do, and this gap can be emotionally uncomfortable. In order to leverage the tension resulting from this gap in a positive way, we need to understand our emotions better and refine them. In this way we can transform the potentially destructive power of emotions into a positive force for change. The TP tools help people to manage the interdependency of intellect and emotion in the change process. In this way, change can become the transformational effort that is required for any sustainable, as opposed to temporary, shift.

10.5 Shifting Beyond the Hierarchical Mindset

The major problem with replacing a hierarchical mindset lies in the subliminal, unchallenged mental models that make us believe that an organization requires a superimposed control mechanism, be it a boss, a function, or an accounting structure based on cost accounting type considerations. The TP tools help us understand the connections, linkages, and the overall mechanism by which we infer reality. Reality is shaped in our minds by connections that largely remain unchallenged unless we unveil them. By making explicit the cause-and-effect relationships with which we perceive reality we have an opportunity to challenge all of those assumptions that limit our ability, for instance, to work within a non-silo infrastructure, meaning within a network of projects. More in general, the good functioning of an organization where conventional hierarchy has been challenged lies in enabling higher forms of "thinking systems" that are stifled both by current educational systems organized into silos and the corporate world.

It is not enough simply to change processes or make organizations flatter. Flat organizations could easily turn into a short-lived gimmick unless we *systemically challenge* the working and, ultimately, existential paradigms that govern the image that we have of what it means to live and work together. This requires a considerable cognitive effort and without the right cognitive tools to support that effort, there can be little guarantee of continued success. The next section is dedicated to explain how the Thinking Process Tools from the Theory of Constraints can do precisely that.

10.6 The Pattern for Change with the Thinking Process Tools

In the Decalogue approach, the TP tools not only support the change process, but also they are the tools we use to *implement actions* at every level. All the conversations for action that take place in the network can be transmitted through the use of these tools, thus focusing the thought processes involved, accelerating discussion and consensus, and providing easy to communicate, visual and verbal results. These are powerful enhancers of involvement and flow. Through continued and ongoing use, they can reduce the variation that inevitably arises when points of view and ways of expressing those points of view are allowed to deteriorate, thus increasing quality even at the level of human interaction. They can be used in a continuous cycle, from creating overall long-term strategy, down to the daily guidelines for carrying out repetitive tasks and new tasks alike. They provide a tangible resource to reinforce the systemic nature of the work, and having such a resource is crucial to support the cognitive challenges involved.

By following the logical sequence in the use of the TP tools, as illustrated below, organizations can shift away from an organizational model that is inhibiting their full potential toward a model based on a network of synchronized projects using the Critical Chain algorithm. Indeed, the journey starts with the building of a "core conflict" (Figure 10.1).

10.7 Building the Core Conflict Cloud

Using cause-and-effect logic, the building of a core conflict provides us with a cognitive snapshot of the current reality of an organization. This is achieved through gathering our intuition and categorizing it into five categories: undesired reality, desired reality, needs underpinning those positions, common goal, and assumptions (mental models). This action alone is a huge leap forward in our understanding of our situation.

We start by identifying the Undesirable Effects (UDEs) that the organization is experiencing in its way of working. This is a list of factors that are creating discomfort and represents the "symptoms" of an underlying root cause. We may even describe them as *emergent properties* of the network of interactions within the organization. Once identified, these UDEs are summarized into one, overarching UDE that sums up the current, unsatisfactory reality. We then verbalize what a desirable reality would look like. We call these positions in the conflict cloud D and D′. These become the two conflicting positions D and D′ of the "core conflict cloud" (Figure 10.2).

This allows us to move on and identify the profound needs underlying these conflicting positions that drive an organization. We state the need for "control" that forces us to accept or to cope with D; we call it B. Then we

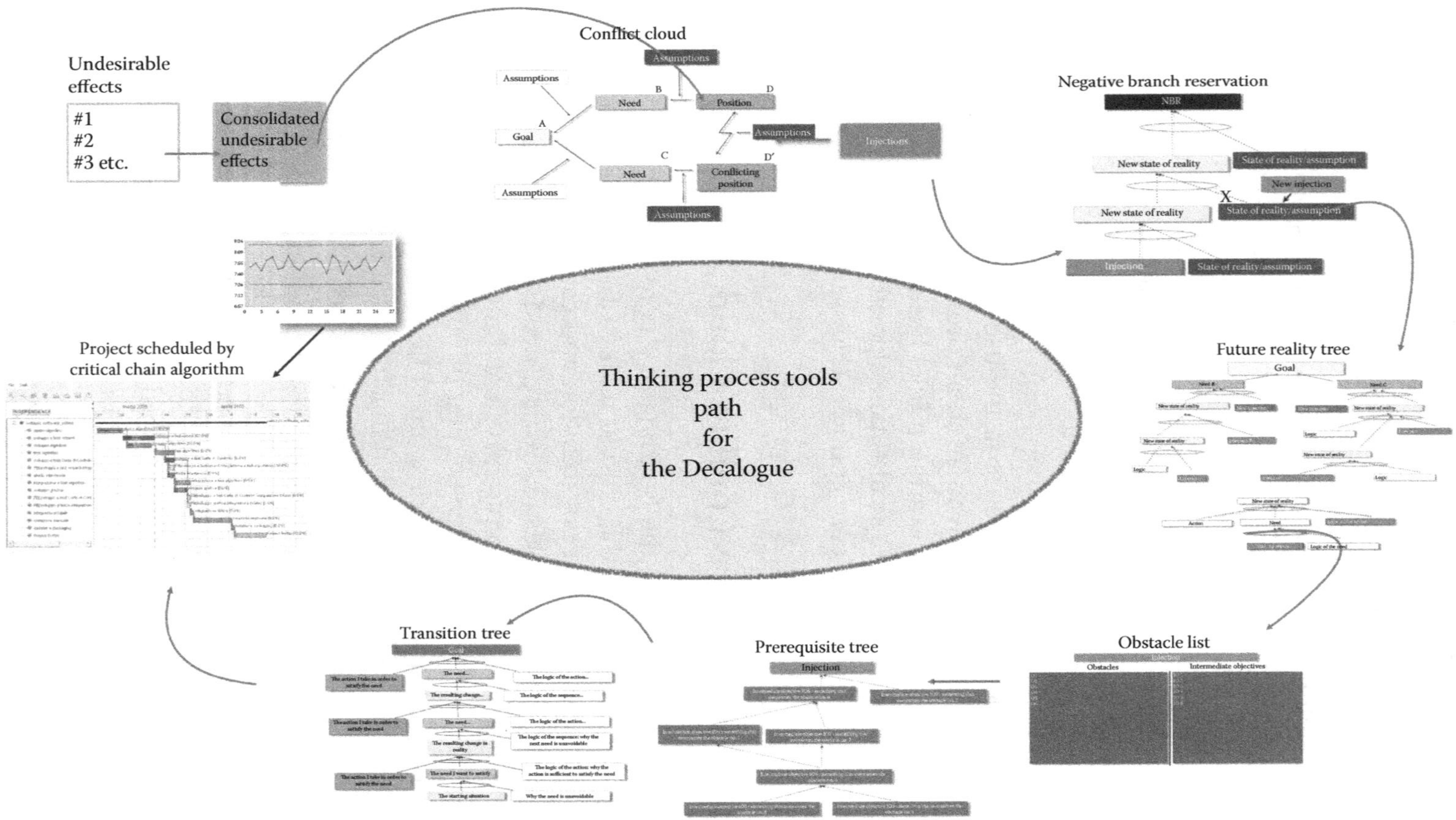

Figure 10.1 Complete cycle of thinking process tools used in the Decalogue approach to transformation.

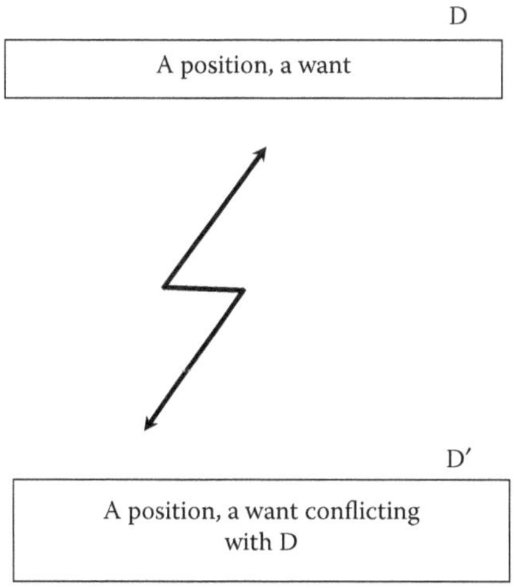

Figure 10.2 Positions D and D′ in the core conflict.

state the need for "vision" that prompts us to say that D′ is the reality we would like to live in; we call it C (Figure 10.3).

Once these needs are precisely verbalized, the organization can then derive the common goal that satisfies those needs, thus providing an organic direction rather than artificially imposing a goal (A). The needs B and C must be simultaneously satisfied in order to achieve the goal in A (Figure 10.4).

The first phase of building a core conflict generally involves a group of leaders and managers over a couple of days in a room who are free to "bitch and moan" until all the UDEs are verbalized. This first phase is a very "feel good" one where everybody agrees that the company is plagued by these effects. These effects are and feel "real" and everybody would like to get rid of them. Summarizing all the UDEs in one single statement is normally a

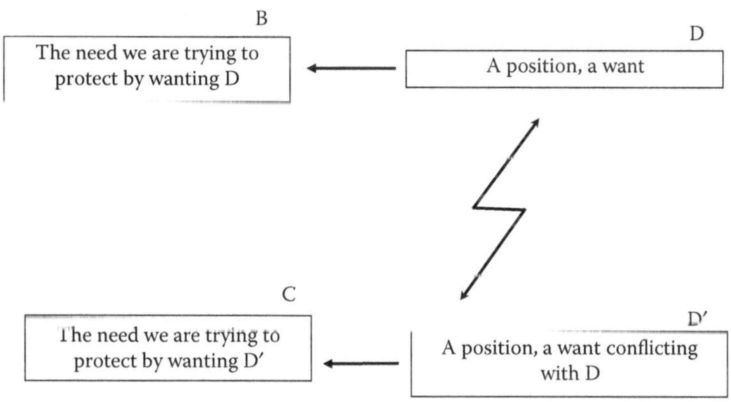

Figure 10.3 Needs B and C.

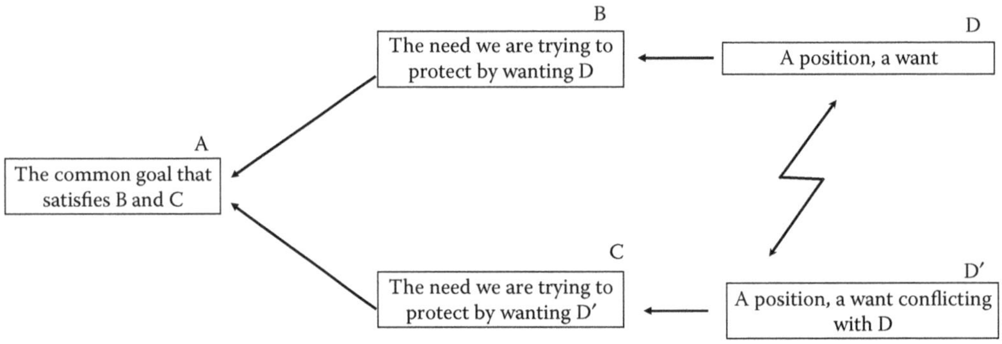

Figure 10.4 Core conflict cloud with positions, needs, and goal.

little cumbersome but it is generally done in a few hours. At this point, the procedure of identifying the desired reality, needs and goal begins, and the end result is normally welcomed as a breakthrough. How does that breakthrough happen?

The conflict cloud helps to sharpen our intuition. In just a few days the group of managers has moved from an often disparate set of nonverbalized hunches to a clear cut picture of the forces that keep them from achieving their goal. Moreover, a precise description of the needs that craft the psyche of the organization goes a long way toward helping to understand the "why" we are trapped in this conflict. We believe that no top management strategic retreat session delivers a tangible and operational output like this one. Now that the intuition is strong we can make it stronger.

What transforms a core conflict into a full-blown picture of our current reality is a disciplined, orderly elucidation of all the mental models that give birth to the conflict. These mental models are deeply rooted images that we have of ourselves and the world around us. These mental models, which we may also call "assumptions," are the cognitive lenses through which we perceive reality.

Systematically surfacing the assumptions is the most challenging aspect of the core conflict cloud because we have to think hard to "smoke them out." Assumptions are, like any other mental construction, the result of factors that are external (the environment, education, experiences, values, etc.) and internal (the chemistry and physics of our mind). The difference between an assumption and a statement of reality is only the realm of validity and this is determined often by cultural circumstances. A practical example of this would be to take a sentence like "In a democracy every citizen is entitled to decent, affordable, and reliable healthcare." and ask for a comment from a statistically representative sample of individuals in the United States, Canada, and Europe.

Assumptions are the logical connectors between goal, needs, and wants; they help us see the logic that shapes the conflict. A conflict with its set of clearly verbalized assumptions portrays the current reality precisely in the way we experience it and is the strongest possible support we can provide to our intuition (Figure 10.5).

Goal, needs, wants, and assumptions tell us why we are in the current state of reality. But they also pave the way to come out of this reality and move toward a future that is more desirable.

As we said, assumptions are mental models that we have about the world; they are formed as a result of experiences and sociocultural circumstances. Assumptions are, in every respect, a reality for the person that develops them. These assumptions, particularly the ones that we verbalize between D and D′ in the conflict are, *de facto*, the constraining element of our reality; they are our *cognitive constraint* (Figure 10.6).

Once completed, the core conflict cloud provides us with two essential elements:

1. A root cause analysis of what is keeping the organization stuck
2. A clear verbalization of the goal they desire to accomplish

This corresponds with the first phase of change: "What to change."

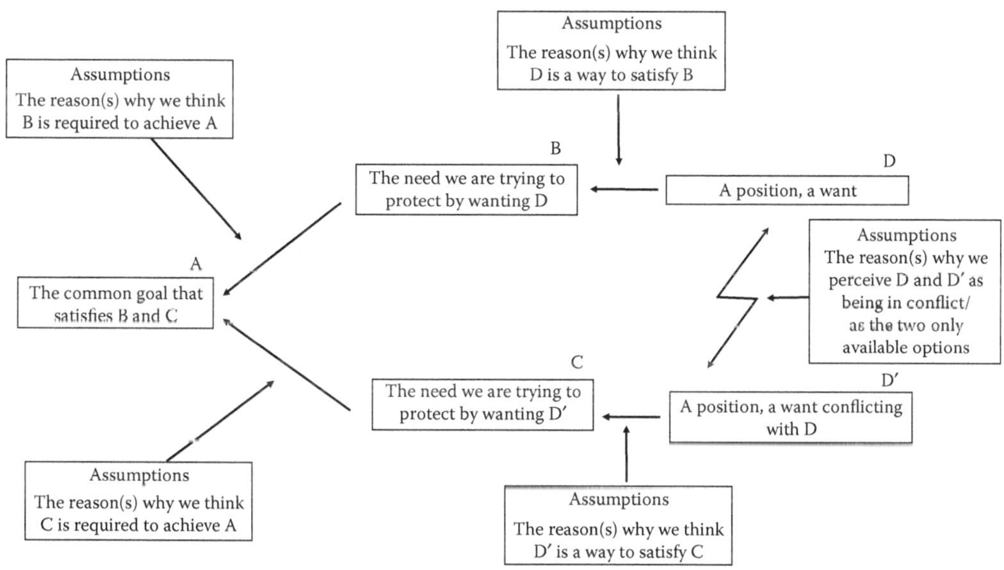

Figure 10.5 Conflict cloud with assumptions.

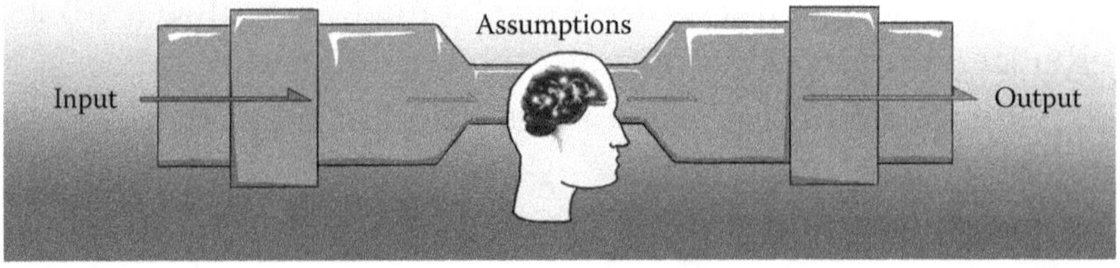

Figure 10.6 Cognitive constraint.

10.8 Finding the Breakthrough Solution with "Injections"

As we said, assumptions are mental models we have about the world; they are formed as a result of experiences and sociocultural circumstances. Assumptions are, in every respect, a reality for the person that develops them. The assumptions that underlie the core conflict are what keep the organization stuck in its current reality. By systematically invalidating these assumptions, it is possible to verbalize "injections" to the conflict, in other words, statements that invalidate the assumptions while respecting the needs verbalized within the conflict. This is where breakthrough solutions can be developed. We use the term "breakthrough" for any comprehensive set of solutions that invalidate assumptions to a conflict; the more "core" is the conflict, the more powerful the solutions must be.

When the constraining beliefs are challenged and invalidated by identifying logical statements that would disprove them, then these "constraints" are "removed." As a result of this removal (the TOC term is "elevation") of the constraint, our ability to achieve our goal is magnified. In order to qualify as "assumption sweepers" these logical statements disproving our assumptions must fulfill two prerequisites:

1. They must logically invalidate one or more assumptions
2. They must protect/address both needs OR one of them and be neutral to the other

Indeed, the totality of these statements must address/protect both needs. If these prerequisites are satisfied, we call these statements "injections." The need for control (B) and the need for vision (C) are captured by the two statements: on one side the "vision" of a company that can overcome with ease the limitations they clearly see as artificial, on the other the "controlling" need for remaining as faithful as possible to the perception they have of themselves professionally and otherwise.

Injections are solutions to the conflict; by invalidating all the assumptions they "evaporate" the conflict cloud (D and D′ disappear) and can potentially move us from our Current Reality to a more desirable, less constraining future reality.

However, in order for this to happen, we have to ensure that this set of injections is both complete and as free as possible from potential negative implication. Only then will we have a full *understanding* of the pattern in front of us. Only then will we have a thorough comprehension of all the potential ramifications of the solutions we identified (the injections).

10.9 The Future Reality Tree

These solutions/injections are then connected together using the future reality tree (FRT) tool. The FRT corresponds with the second phase of change, "What to change to." Using a logic of sufficiency, this tool maps out the solutions in a progressive and integrated pattern toward the goal previously identified in the core conflict cloud.

The FRT ensures the completeness of the set of injections identified, hence providing a conceptually reliable path to the future. The process of building an FRT requires some skill, a bit of experience and a fierce determination. It is in no way an academic exercise nor is it an exercise in conventional logic. Building an FRT is only possible if we have embraced the vision and the method that supports it; the vision is that of a company that takes very seriously its commitment to the future, that sees itself as an ongoing generator of wealth for all its stakeholders and society at large. The method is the orderly, relentless identification of all the cause-and-effect relationships that are likely to shape the future if certain actions are carried out successfully. In this sense, the FRT is similar in nature to the PDSA (plan, do, study, and act) cycle because it prompts a rigorous, scientific investigation of the subject matter.

10.10 A Real-Life Example from Core Conflict to Future Reality Tree

The core conflict cloud below portrays the current (at that time) reality of a manufacturing company producing special equipment for the printing industry (Figure 10.7).

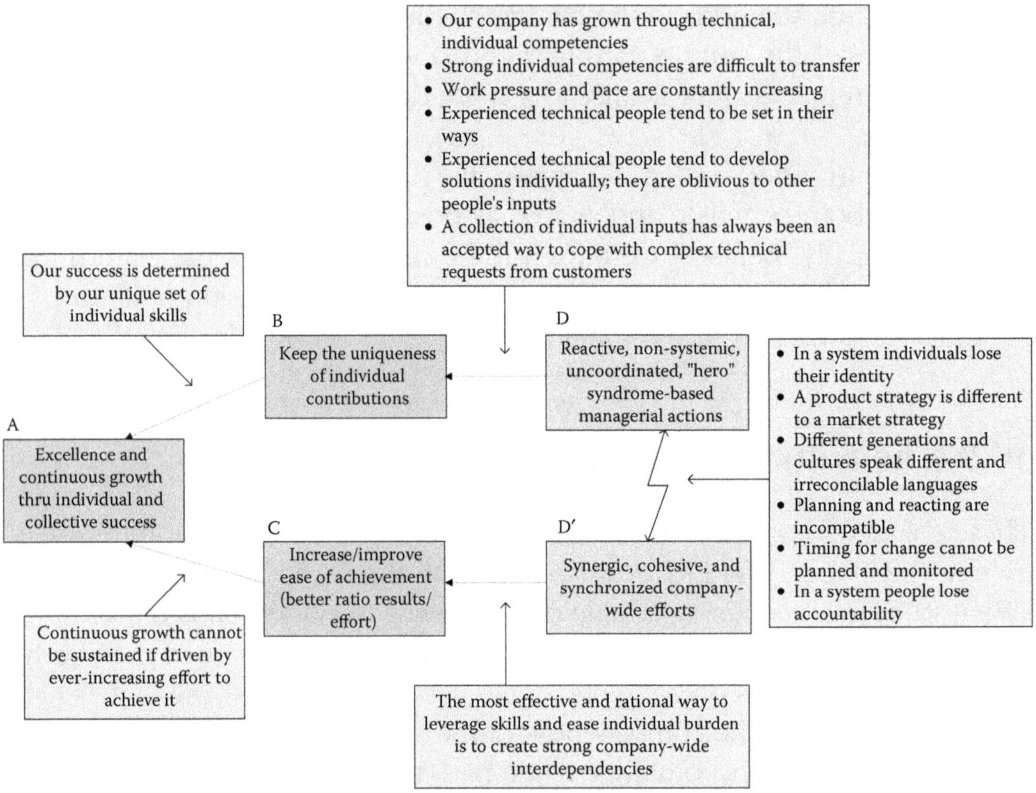

Figure 10.7 ProPrint conflict cloud.

The way to read this conflict is the following: "If our goal is (A) *Excellence and continuous growth through individual and collective success*, THEN we must (B) *Keep the uniqueness of individual contributions*; IF we must (B) *Keep the uniqueness of individual contributions* THEN/THEREFORE we are forced to cope with/accept (D) *Reactive, non-systemic, uncoordinated "hero syndrome"-based managerial actions.* On the other hand, IF our goal is (A) *Excellence and continuous growth through individual and collective success* THEN we must (C) *Increase/improve ease of achievement (better ratio results/effort)*; IF we must (C) *Increase/improve ease of achievement (better ratio results/effort)* THEN/THEREFORE we want/desire (D′) *Synergic, cohesive, and synchronized company-wide efforts.*"

The boxes between A, B, and D; A, C, and D′ as well as the one between D and D′ are the assumptions and they read as follows: "If our goal is A AND Box with assumptions, then we must B (or C)." The same reading applies to the other boxes with assumptions. The Box between D and D′

reads as follows: "We would like to have D′ but we are forced to live/cope with D BECAUSE…"

Here is how we read this conflict cloud in full including the assumptions: "IF our goal is (A) *Excellence and continuous growth through individual and collective success*, AND (assumptions A–B). *Our success is determined by our unique set of individual skills*, THEN we must (B) *Keep the uniqueness of individual contributions*; IF we must (B) *Keep the uniqueness of individual contributions* AND (assumptions B–D):

Our company has grown through technical, individual competencies;
Strong individual competencies are difficult to transfer;
Work pressure and pace are constantly increasing;
Experienced technical people tend to be set in their ways;
Experienced technical people tend to develop solutions individually that are oblivious to other people's inputs;
A collection of individual inputs has always been an accepted way to cope with complex technical requests from customers;

THEN/THEREFORE we are forced to cope with/accept (D) *Reactive, non-systemic, uncoordinated "hero syndrome"-based managerial actions.* On the other hand, IF our goal is (A) *Excellence and continuous growth through individual and collective success* AND (assumptions A–C) *Continuous growth cannot be sustained if driven by ever-increasing effort to achieve it, THEN* we must (C) *Increase/improve ease of achievement (better ratio results/effort)*; IF we must (C) *Increase/improve ease of achievement (better ratio results/effort)* AND (assumption C–D′) *The most effective and rational way to leverage skills and ease individual burden is to create strong company-wide interdependencies,* THEN/THEREFORE we want/desire (D′) *Synergic, cohesive and synchronized company-wide efforts.*" We would like to have D′ but we are forced to live/cope with D BECAUSE we make these assumptions between D and D′:

- In a system individuals lose their identity
- A product strategy is different to a market strategy
- Different generations and cultures speak different and irreconcilable languages
- Planning and reacting are incompatible
- Timing for change cannot be planned and monitored
- In a system people lose accountability

In our experience no person or company has ever benefitted from reading another person or company's core conflict. The reason for this is multifaceted. First, every company believes it is special and unique—nobody compares; moreover it is difficult to penetrate quickly somebody else's subject matter. Second, reading a conflict is not the same as building it; the focus and comprehension in the two scenarios are radically different. Third, the language that management uses to portray its reality is peculiar to that group of people. This is why, in fairness, we do not believe reading this cloud and what follows will be of great help. However, in order to increase even minimally the chances of being successful, we have done some editing work to the original conflict so as to make it as general as possible.

The company, which we shall call ProPrint, had about 200 employees and two facilities. It had been on the market for 50 years and had been bought out by some of the existing management who had teamed up six years previously with some keen, willing and long-term focused capital. ProPrint was well known for the quality of its products, had a well-established customer base, no union issues, and enjoyed a respectable cash profit generation.

The main UDEs of ProPrint were all connected with its legacy. It had been a small, family-owned company for decades, where hands on experience and closeness to the owner/founder were valued more than formal education and knowledge. The skills considered critical for the product were in the secretive hands of a few elderly employees, and the focus of the company was exclusively on the product rather than the customer. The sudden jump in revenue ensuing the takeover by management had created a clear disconnect between traditional work habits and what was needed to cope with the pace of growth. Moreover, ProPrint did not seem to be able to take advantage of a growing demand coming from abroad.

In the course of a three-day session, the top management team of eight people were able to clarify and fortify their intuition about the issues they were facing, that is, the conflict that was trapping them. They were able to verbalize clearly the forces that were keeping ProPrint from evolving. The management at ProPrint invested heavily toward this end and they came out with the considerable result we reproduce here. We took the liberty of making some simplifications to make it more legible (Figure 10.8).

This is the part of the FRT that connects the injections to the achievement of the goal via the simultaneous satisfaction of both needs. (A more complete and, for the purpose of this book, unnecessary version of this tree should show the links between the achievement of the goal and all the DEs that would replace the UDEs out of which the core conflict was developed.)

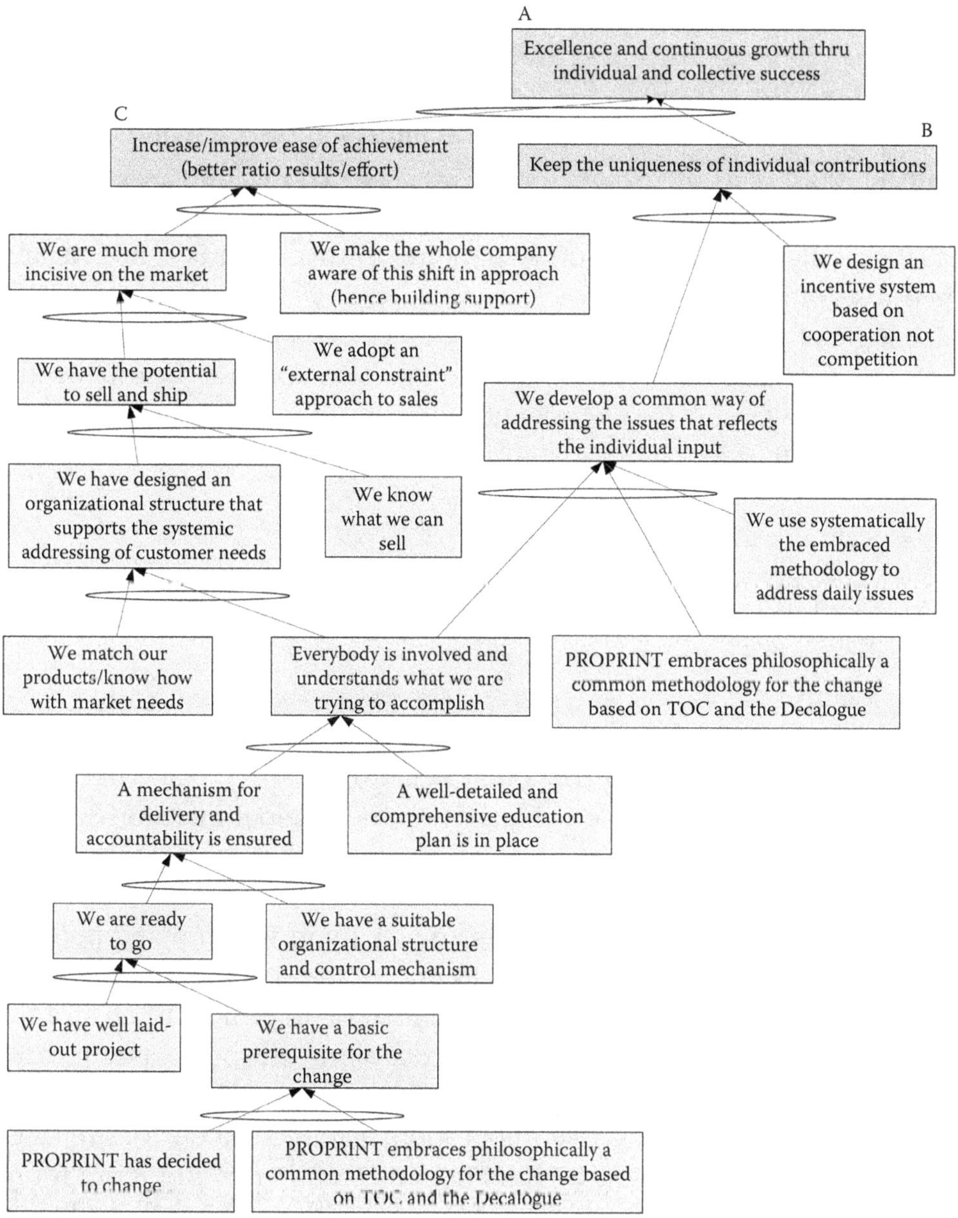

Figure 10.8 ProPrint future reality tree.

The set of injections developed by ProPrint was validated in terms of their completeness by linking each injection through cause-and-effect logic with "future" intermediate states of reality that are achieved as a result of the accomplishment of the injection. Indeed, these injections are "pies in the sky"; they do not exist in reality and they are, at this stage, only logical statements.

The strength and the value of the FRT are in the understanding it provides of the comprehensiveness and the breadth of the effort needed to transition from the current to the future state of reality. The thoroughness of this effort and the completeness of the understanding we derive from it are further enhanced by the meticulous test we carry out on potential negative implications deriving from the coming into existence of the injections. The TP tool that supports our understanding in this effort is called negative branch reservation (NBR).

The logic of this tool is simple but the mechanics require some attention. In essence, with this tool we verify if the injection suggested carries with it potential negativities; if we can identify them we can try to diffuse or reduce their impact. The steps to build this "branch" are the following:

1. We state the injection
2. We state the potential negative outcome
3. We build the cause-and-effect chain of events that we anticipate would determine the negative outcome
4. We identify the logical statement/statement of future reality that would turn the positive of the injection into a negative
5. We devise an action that trims that negative
6. We incorporate the needed change into the original injection so as to make it more effective

Anticipating negative implications is an activity in which many people are well versed. The point here is not to encourage skepticism, fuel negativity or stifle creativity and enthusiasm. The goal of an NBR is to ensure full understanding of the ramifications of our decisions. In some cases this tool helps us abort in time a half-baked decision. In some others, regardless of our inability to trim that negative implication, it reinforces our desire to go ahead anyway. More frequently, NBRs help us craft better and more rounded injections giving us a higher chance of accomplishing what they had been designed for.

To exemplify the use of the NBR we can see how ProPrint raised and trimmed a potential negative associated to one of their injections on the FRT (Figures 10.9 and 10.10). (The injection is in the bottom left box.)

So far, we have worked with the basic starting situation and pictured it using the tree of our current reality. We have seen how ProPrint built its own current reality. Depicting their reality so clearly helped the management team to develop the focus and the desire to build a thorough understanding

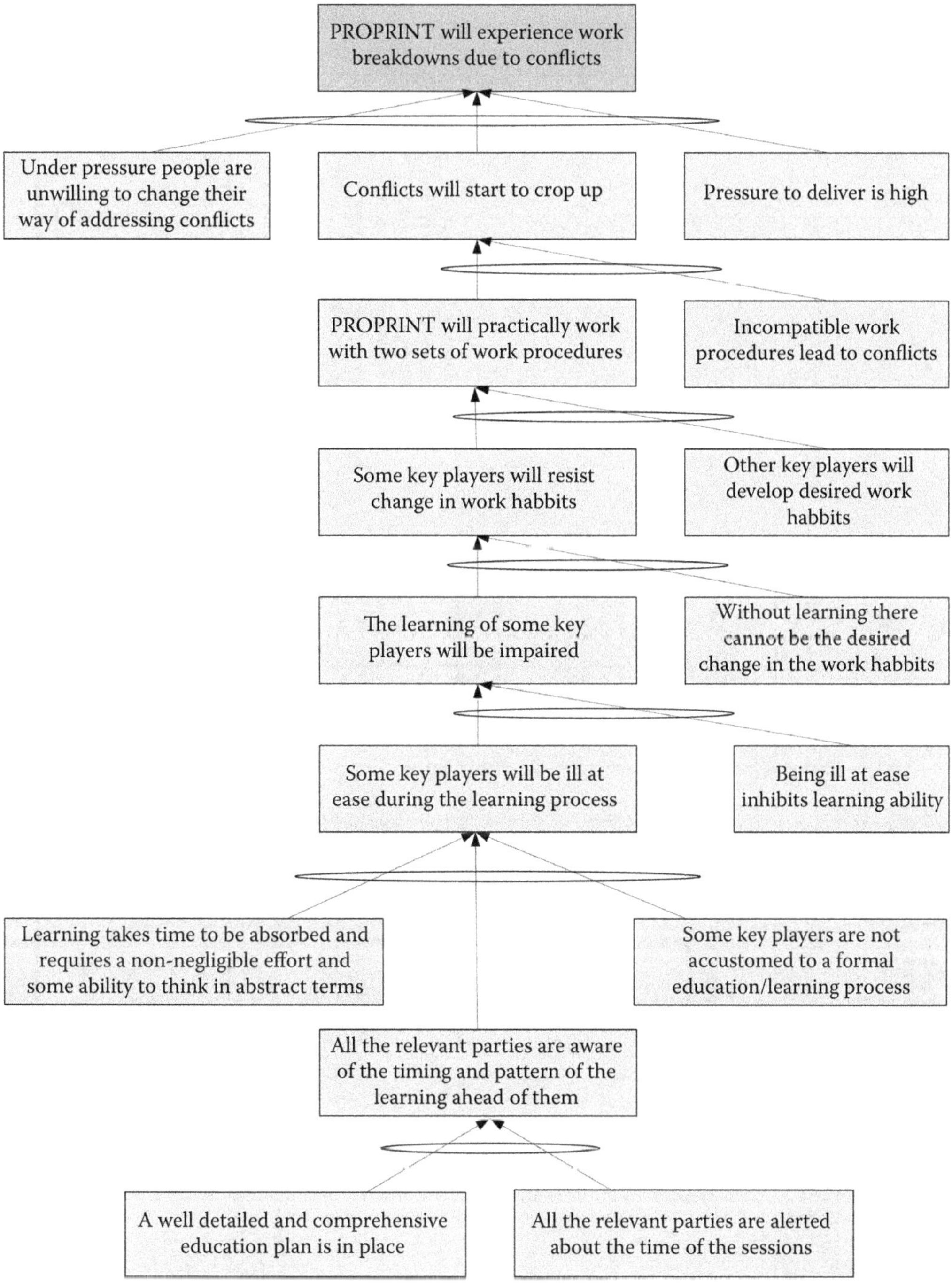

Figure 10.9 ProPrint negative branch.

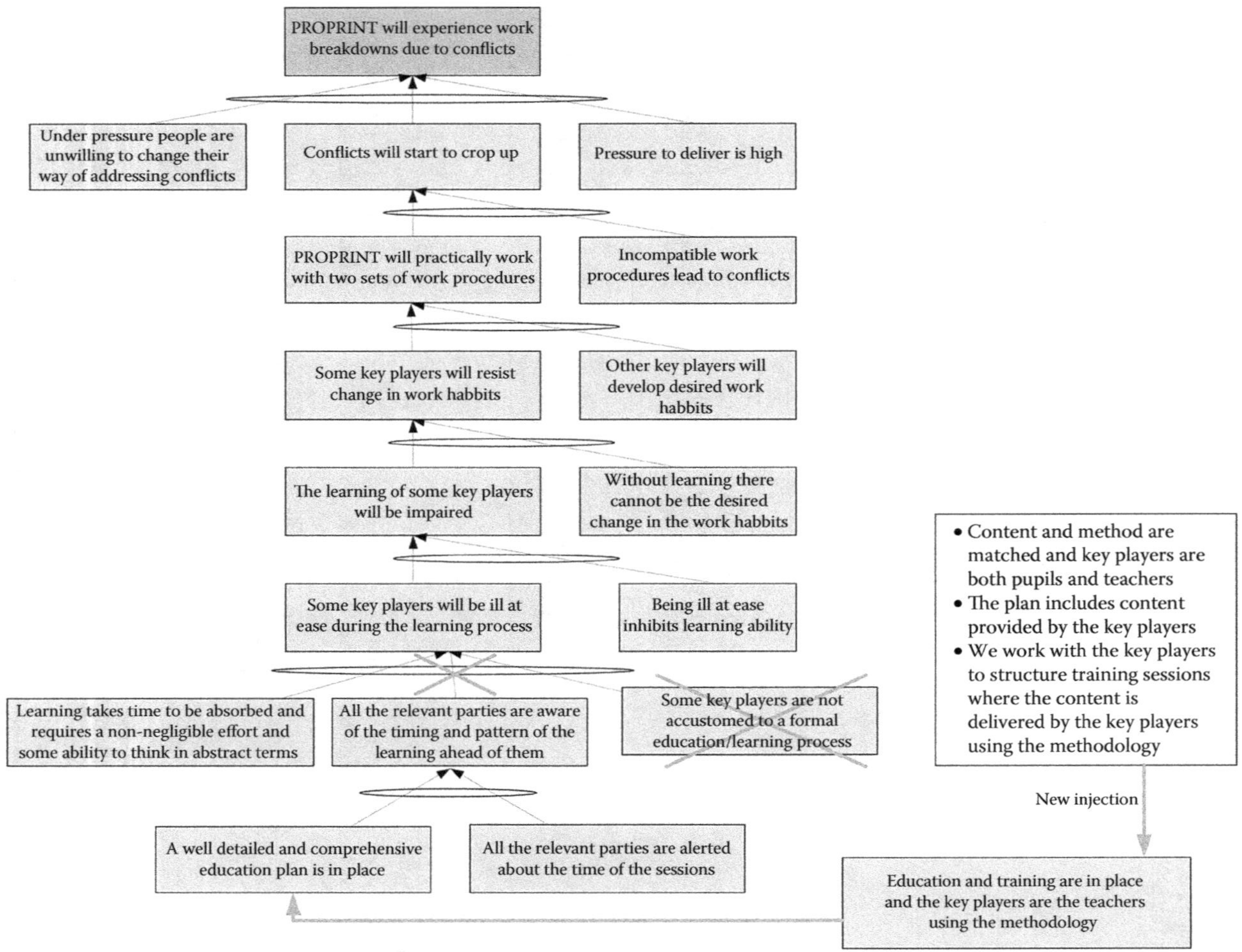

Figure 10.10 How ProPrint Negative branch is trimmed.

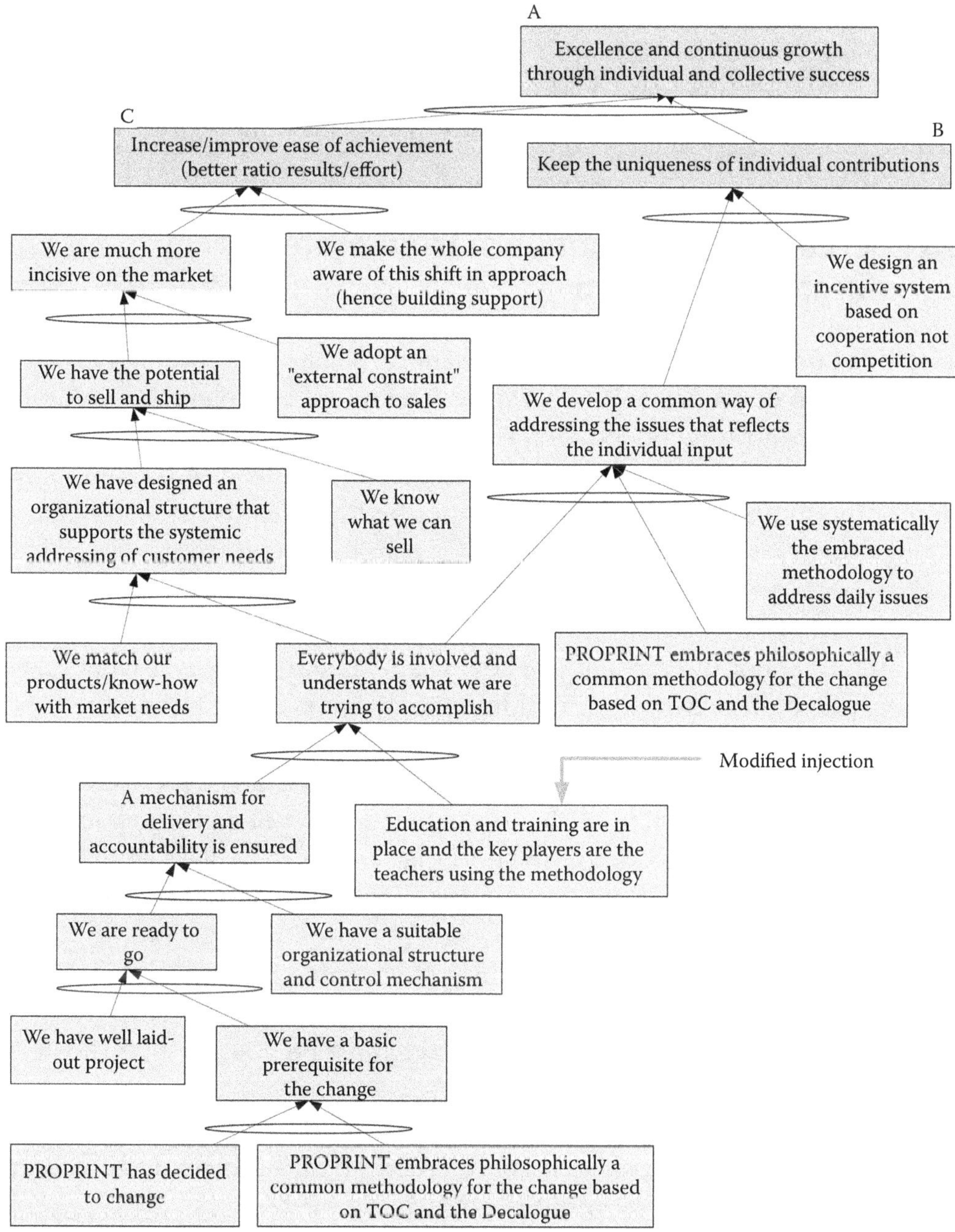

Figure 10.11 ProPrint FRT with modified injection.

of how it would be possible to evolve out of that reality toward a better future. The FRT and the NBRs provided that understanding and the global vision of how the company could move ahead without losing itself. The bridge between the present and the future had been built (Figure 10.11).

10.11 Creating the Continuum from Systemic Strategy to Action

Let's allow ourselves a moment of speculation. What is the future? This is not a rhetorical question, on the contrary. The life of an enterprise, its ability to stand the test of time, its possibility to wade through the ups and downs of the markets rests firmly on its ability to pursue its vision of the future. A vision of the future for an enterprise cannot be the outcome of a strategic meeting where a few executives develop a plan to react to some ill-understood events or to push something based on personal agendas. Nor can it be some opportunistic decision driven by purely monetary reasons.

The future is not the present in a few years' time, neither is it what happens regardless of what we do. Building the future is about taking the responsibility to make the future happen; it is about striving to understand better and better the implications of what we do. A vision of the future is certainly the responsibility of the top management but calls, first and foremost, for the ability to subordinate to that vision, regardless of the circumstances. People's *involvement in building a future they can associate themselves* with is essential.

If we take our understanding seriously, if we value our ability to anticipate what future events will be like as a result of our actions, if we truly believe in cause-and-effect thinking, then we must be able to subordinate to that future reality and walk on that bridge.

The bridge that brings us to the desired future and that makes that future possible is *knowledge*. Knowledge is not erudition and it is not information. Knowledge is not knowhow and it is not experience.

Knowledge is based on theory and must allow prediction. Without the ability to predict there is no management. However, in order for this knowledge to deliver results we must be one with it. Knowledge cannot be disconnected from the consciousness that should permeate it. If the knowledge needed to execute the injections of the FRT is not in synch with the change in reality that the application of that knowledge will create; if the unfolding of the pattern we have chosen creates a disconnection between the holder of the knowledge and the sense of who they are in the new reality; if what people know does not match the image they have of themselves in the new reality, then the transformation from the present to the future will not happen.

There are two TP tools that help to deploy the necessary knowledge and the possession of it together with an increased self-consciousness: the prerequisite tree (PRT) and the transition tree (TRT).

10.12 Gathering and Deploying Knowledge to Reach the Goal

The PRT and the TRT have been designed to gather and deploy all the knowledge needed to achieve the injections in the FRT. The PRT and TRT are the backbones of an actionable plan. They pursue the goal of providing guidance and a logical pattern as well as detailing all the steps that have to be taken in order to achieve the stated objective. Let's see how they work.

When we face the ordeal of achieving an ambitious target we are often daunted by the "mountain" we have to climb; all we see are the obstacles in front of us. We perceive these obstacles as undefined, scary entities that hinder our ability to achieve our goal. When we are in front of an injection (or any other arduous task), the first reaction is "Oh my goodness, how am I going to do this?" This is understandable but it is not rational. First of all, why are we faced with this feat? Because either we have chosen to be, or because it is part of an evolution, personal, and/or organizational, we have decided to be part of. Very rarely do we get ourselves in front of a task we have no possibility of overcoming. Indeed, at the onset, we might not have clear ideas on how to tackle the ordeal. This is good and we can leverage it.

The basic raw materials to build a core conflict are UDEs; the starting ingredients for a PRT are the obstacles we envisage in our journey to the goal: let's list them. The writing of this list is going to be both liberating and informative: we shall have full clarity of what we perceive as an impediment and with it we'll develop a pacifying sense of control over it. When the list is complete we will have broken down "a big and undefined problem" into a set of much smaller ones. How do we attack them? We make a list of the obstacles on the left side of the page, from top to bottom; then on the same line on the right side of the page we can list a corresponding Intermediate Objective (IO). An IO can be either a logical statement describing what would enable us to overcome that obstacle or, better, how we intend to do it. For instance, if the obstacle is "I have insufficient funds" the IO could either be “I have provided for sufficient funds” or, better, the solution, that is, “I have persuaded my aunt to sponsor me.”

10.13 How to Cause the Change—Obstacles along the Road to Change

In order to build the PRT, we start by listing all the obstacles we see preventing us from implementing the injection (Figure 10.12).

Now that we have the solution (IOs) to the individual obstacles (very often one IO addresses several obstacles) we can forget about the obstacles and concentrate on the IOs. The next step is to provide a basic sequence to these IOs. This is for two reasons: first, we need to ascertain “what is prerequisite to what”; second, we need to know how many things we could theoretically run in parallel. (In reality, resources will dictate that pace, as we will see.)

Then we sequence the IOs, placing at the bottom of the tree the first ones we have to achieve and then adding the others, according to a logic and time sequence. The PRT is based on necessity.

The PRT and the TRT (see below) were both developed for the same purpose but they are “genetically” different. A PRT is a map; it is based on a logic of necessity and does not foresee any check based on conditions of sufficiency. It is, essentially, a roadmap, a guide that provides a suitable route to the goal. Its main value lies in the collaborative effort required to build it and it is an ideal tool for teamwork. It is the simplest of the TP tools to learn but it can be misleading; it is not just a list, it is a list with priorities dictated by logical prerequisites and these prerequisites are, in turn, altered by the amount of resources at hand. You will be surprised to see how many

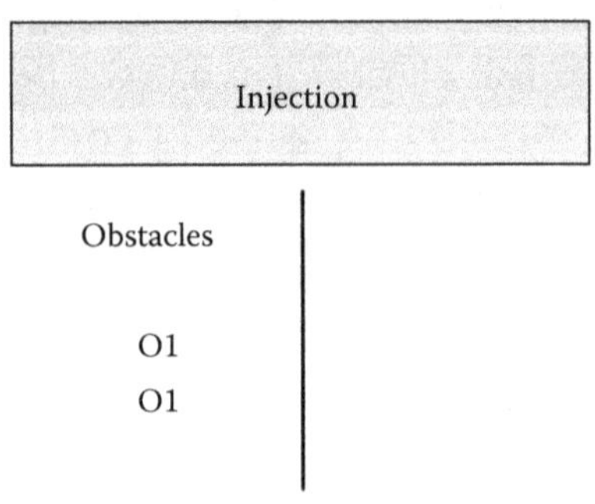

Figure 10.12 Injection and list of obstacles in the prerequisite tree.

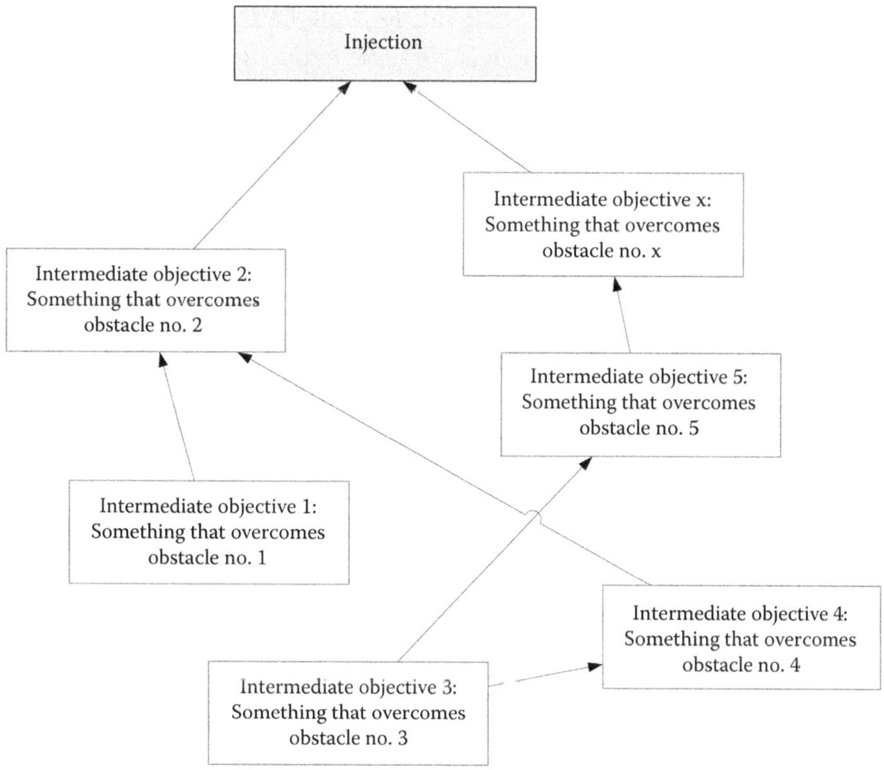

Figure 10.13 Intermediate objectives sequenced toward injection.

different points of view there are in a group trying to address "what is logically a prerequisite to what" (Figure 10.13).

10.14 Making the Change Happen with the Transition Tree

A TRT is a conveyor of understandable, usable, and comprehensive knowledge aimed at executing a precise task with very little variation. Transition trees are developed any time it is necessary to convey to somebody a precise procedure to transform an input into an output. If it were to be used within the framework of a conventional quality assurance documentation, it would certainly be in the so-called "work instructions." TRTs use an extremely powerful cause-and-effect sufficiency logic; in other words, every intermediate output is checked in terms of the sufficiency clauses that determine it. The transition tree is built by stating the final result we want to achieve, normally an IO of a PRT. The basic, repetitive, five-block structure of a transition tree is the following:

(1) A starting situation, (2) the prevailing logic of the situation, (3) the need triggered by the prevailing logic, (4) the action taken to satisfy the need, and (5) the logic of the action—why we have taken that action (Figure 10.14).

As a result of the action taken, the reality will change to a new state and we can easily verify if we are getting closer to the final result we want to achieve with the TRT. At every step, we can scrutinize the effectiveness of our action, the validity of its logic, and their coherence with the identified need.

PRTs and TRTs are the TP tools we use to ensure that all the knowledge available is captured and used in an orderly and proficient manner. Moreover, we can use them as a backbone for any project we want to undertake in as much as they provide a clear-cut set of actions that can be easily scheduled and staffed. In other words, the actions on the TRTs can be easily transformed into tasks of a project with an appropriate duration and allocated resources.

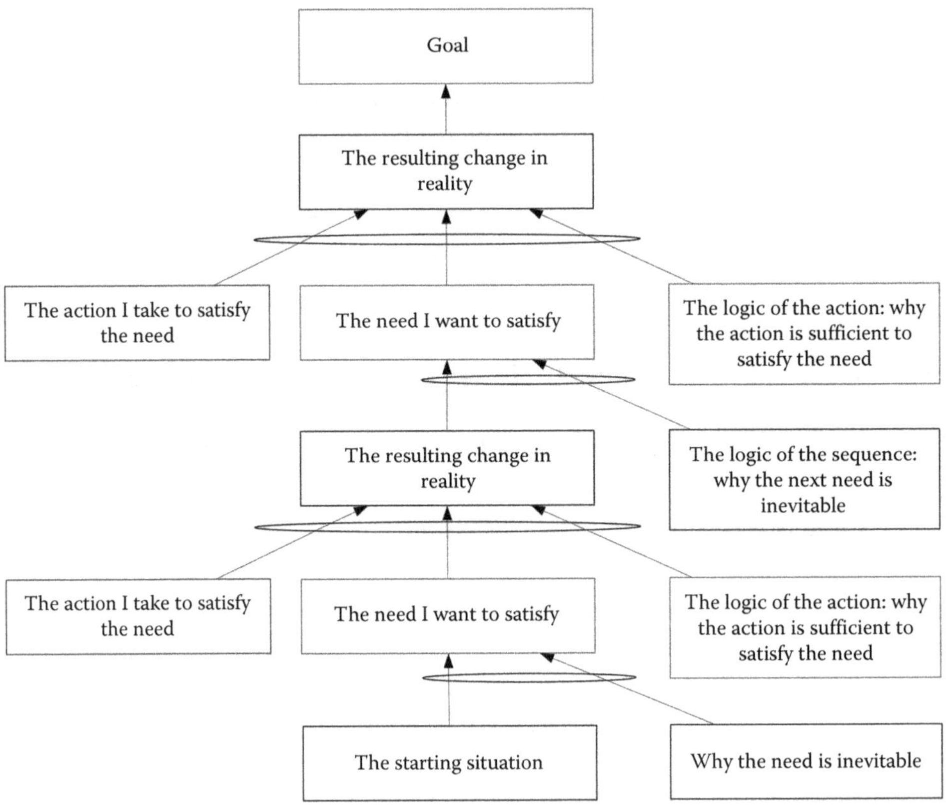

Figure 10.14 Template for a transition tree.

What we have tried to describe is the pattern that links the different phases of a transformational project: what to change—what to change to—how to cause the change. These phases activate very distinct elements of our intellect: intuition, understanding, and knowledge, which the human mind struggles to hold together with the cohesiveness required for a minimally complex project.

The TP tools have been devised to improve this cohesiveness and help us access a much higher level of *systemic intelligence*. These tools are the quintessential support to systems thinking and, just as with SPC, it will take humankind several decades to appreciate their depth and scope.

ProPrint is a success story not because of their ability to execute the FRT and enact successfully all the trees needed to accomplish the stated goal. It is a success because it provided a vibrant and ambitious management team with the possibility to gauge their strengths and weaknesses in respect with their legitimate ambitions. As a result of this exercise, they realized that the effort required to truly transform successfully the organization they had taken over was beyond the desire they thought they had. They realized that what they had achieved could have been very easily (and very rewardingly) incorporated by a larger entity and become a part of a bigger organization. Accordingly, they accepted a very generous offer from a large multinational to become one of their subsidiaries, performing better and better what they already knew how to do. Everybody won.

10.15 To Summarize: The Cycle of the Thinking Process Tools in the Decalogue

By following the logical sequence in the use of the Thinking Process Tools, as illustrated above, organizations can shift away from an organizational model that is inhibiting their full potential toward a model based on a network of synchronized projects using the Critical Chain algorithm. The cycle of tools moves an organization through the three phases of change, "What to Change," "What to Change To," and "How to Make the Change Happen."

What to Change:

- Identify the "UDEs" of the organization (symptoms of underlying root cause)
- Summarize into one, overarching UDE that sums up the current, unsatisfactory reality

- Verbalize what a desirable reality would look like so we now have the two conflicting position of the "conflict cloud"
- Identify the profound needs underlying these conflicting positions (control and vision)
- Surface systematically all the underlying assumptions (mental models) that connect the statements contained in the conflict

What to Change To:
- Systematically invalidate the assumptions between D and D′ to achieve "injections" (systemic solutions)
- Connect injections together using the FRT tool

How to Make the Change Happen:
- Build a PRT for every injection
- Build a TRT for each IO (where needed) to identify tasks
- Schedule these tasks can then be scheduled into a project using Critical Chain
- Repeat: the cycle of tools can be repeated on an ongoing basis, thus conforming to the plan–do–study–act cycle of continuous improvement advocated by Dr. Deming

10.16 Understanding, Knowledge, and Science in a Conscious Organization

Accomplishing the organizational transformation required for all organizations to fulfill their role of sustainable wealth creation does call for a higher and better use of our intellect. The thinking leveraged by the tools designed by Dr. Goldratt is critical to this end. Understanding and knowledge must be solidly linked to the scientific approach and the wealth of discoveries it has generated in the last 350 years and that it continues to generate.

The essence of this scientific approach is embodied in the plan–do–study–act, PDSA cycle, advocated by Dr. Deming as the main mechanism to generate and sustain the application of knowledge within any organization. Each of the four steps of this cycle must be guided by statistical insight. The PDSA cycle is rooted in the epistemological belief that phenomena must be described and understood in statistical terms; this statistical vision of the world, largely applied in the investigation of the natural world, has been adopted by organizations in a limited way and completely misapplied or ignored by economists and financiers, not to mention accountants.

An organization with a high level of consciousness as we have described has, by definition, a high level of interconnection with those who work within its system, with those who supply it and those who are its end-users. The ongoing task is to constantly satisfy better the needs for control and vision, of those who work within it, and of those who interact with it up and downstream. The inevitable outcome of this level of consciousness and connection is a more *ethical* way of operating. There is automatically no space or use for hedging and undercutting, or for pollution of the environment. The inevitable outcome is enterprise with a conscience.

Our ability to create and follow this kind of organization is directly linked with our ability to increasingly *do who we are.* This requires us to accept a level of personal freedom to which few are accustomed and even fewer feel comfortable with. This does not mean laissez-faire, it does not mean just doing whatever you want. It means accepting the responsibility of understanding that our only true limits are ourselves and what we are able to perceive for ourselves. Our limits are mental models, and our mental models dictate the boundaries of our actions. As we begin to challenge these models/assumptions, we begin to taste the vivifying experience of exploring our true potential. Fulfilling that potential is not a question of luck but one of choice.

The prevailing management style (or lack of it) has taken the separation between knowledge and consciousness to an extreme and impaired people's ability to choose intelligence over stupidity. This ability needs to be given back and enhanced manifold. Organizations must overcome the damage caused by this disconnection by leveraging this intrinsic unity between consciousness and knowledge, by relearning how to connect learning and choice and by retooling its people's ability to manage intelligently. A new covenant is required, one where each individual has the opportunity to better themselves and, in so doing, better their organization. We have the intuition, we have the understanding and we have the tools to make it happen. However, in order for the transformation to truly take hold we need to create a new kind of leadership and that leadership must be the expression of a new economics. We will look at these aspects in the next chapter.

Chapter 11

New Intelligence—New Leadership—New Economics

> ...the job of a leader is to accomplish transformation of his organization.
>
> **W. Edwards Deming**
> *The New Economics*

In this book, we are working toward much more than improvement. We are highlighting a path for transformation. To transform what? The prevailing silo-based management style into one of system optimization. This transformation requires relentless effort, but ultimately, it allows organizations to consistently fulfill their role in the world with the highest quality and speed. This requires knowledge, method, and tools, and it requires a special kind of leader. We are going to take a look at the qualities and characteristics of the kind of leader who can drive the transformation. Later in the book, we will look at the systemic tools that every leader needs.

11.1 Role of the Leader

Anyone who has any experience with corporate life cannot fail to see how profoundly damaged this life can be. The kind of interaction that creates a cage for CEOs and boards is often surreal and sadly too often tragic for all the *stake*holders. Just to make sure we are clear: a CEO spends a lot of their time trying to prove to the board and the investors they represent that what the company is doing makes sense; in order to do so, they meet with the board 3–4 times a year, and engage in conversations that span from strategy to GAAP-based financial results. Indeed, boards are only privy to what the CEO (and the CFO) want them to know; the financial transparency is guaranteed by some audit subcommittee and the strategy is discussed by people that, more often than not, have mental models that are the reason for which they have been summoned. Doesn't that sound a bit like Enron?

> The role of a leader is the transformation of the prevailing style of management into one of (whole) system optimization.
>
> **W. Edwards Deming**

It doesn't have to be like that. As we saw in the previous chapter, an organization can use the TP tools to capture their specific core conflict, that is, the inherent cloud of that particular organization. Once that has been captured and the assumptions are surfaced, we have the possibility to develop a full, systemic set of "injections" to those assumptions. Another way of saying it is: we now see clearly what is keeping us stuck and, logically, we know also how to come out of the "swamp" we are in.

The path out of the swamp, or any path forward, must be sustained by a clearly elucidated cause–effect map, one that inserts some epistemological rigor, some ability to predict the outcome of the actions to be taken and, of course, a realistic schedule of how and when these actions will be taken. This is what we have labeled as "projects."

Intrinsically, the work of any organization is made up of the network of all the projects that emerge from the injections to a well-articulated core conflict. The role of leadership (board and C-suite) in this context is therefore very clear. It is to create the environment for these projects to be conducted successfully; to supervise their execution, to intervene in a timely manner with corrective actions triggered by unanticipated statistical fluctuations and to ensure that every stakeholder benefits from the success of the projects. If somebody is willing to spell out the basic (very few) details for a board to operate in this way, we will be able to spare ourselves much of the chatter on corporate governance that currently clouds the issue.

This transformation is neither easy nor for the fainthearted and it is only possible if we recognize and embrace the new paradigm of interconnectedness we live in. Recognizing the emergence of a new paradigm is very difficult and takes time; but it also takes a new kind of intelligence, one that is capable of understanding the multidimensionality of the implications of our choices: physical, mental, and spiritual.

It is not just about seeing the forest where everyone else only sees trees, it is about understanding the forest as the *emergent property* that is generated when a number of trees grow together.

11.2 New Leadership Methods and Tools

The methods and tools of the new intelligence-based leadership cannot be separated from its systemic essence; they need to reflect its original thought and be its offspring. These methods and tools must be one with the leader; they can have a technical content but cannot be techniques, they must be a direct emanation of a way of thinking.

When presented with some sort of "quality improvement toolkit," Dr. Deming was almost invariably dismissive as he perceived them as either unnecessary or conceptually wrong. Besides a tepid endorsement of the Ishikawa diagram for cause–effect analysis, Deming only relied on predictive statistics studies (SPC) and its natural companion, the process improvement chart, at the time simply called control chart.

Why is process improvement and managing variation through SPC so important? Leaders and managers need to have the ability to make rational decisions about their organizations, and the work of their organizations consists of processes. Therefore, managers should know how processes behave, both now and in the future. *Whether they are aware of it or not, all processes are affected by variation.*

Why should a manager go to the trouble of understanding the statistical behavior of any process? To understand this fully we have to make a distinction between forecasting and predicting. Managers may be asked to make forecasts, and these are empirical suppositions based on previous evidence and, in many cases, fears, and hopes. The usefulness of such forecasts is limited and reliance on them perpetrates a totally inadequate empirical approach to management. Managers, instead, need to be able to predict how any given process will behave in the future as a result of their decisions. This prediction must be grounded in statistical understanding and the epistemological stance it portrays. *Statistical predictability* means that the process oscillates "predictably" within its upper and lower limits of variations. This is the only kind of prediction that leaders and managers can reasonably make and the only one that gives rational value to their decisions, as opposed to management based on hunches.

SPC and process improvement charts are indeed the very backbone of any systems building exercise as they provide insight, not just numbers. (See Step 3 of the Decalogue in Chapter 8.) SPC shelters us from risk of "linear thinking" and avoids the mental trap of spreadsheet-fueled deterministic numbering. Indeed, when properly absorbed and used, SPC is never a technique; it is, as Dr. Deming has said, a way of thinking.

11.3 Leading with Critical Chain

The statistical predictability of the processes making up an organizational system facilitates greatly the creation of a network of projects. We can create this network by leveraging the power of the finite capacity scheduling underpinned by the Critical Chain approach to managing projects.

Critical Chain represents the embodiment of a vision of the organization based on pace of flow, people's involvement, and great emphasis on quality. Quality, involvement, and flow are the basic philosophical pillars of the systemic organization so Critical Chain plays a major role in the building of an intrinsically systemic organization.

The future of the organization is in designing organizations not as functional/hierarchical grids that are clearly inadequate due to their inability to fully support, measure, and promote intrinsically cross-functional activities. The solution is to design organizations as a network of interdependent projects that work together to achieve the goal of the organization. The allocation of people to tasks according to their skills and level of competency produces an effective use of the resources available, a more ethical consideration of people's capabilities, and allows us to unleash the power provided by the "intelligent" implementation of project management. Critical Chain enables all this to happen methodologically and practically.

11.4 Leading with the Thinking Process Tools and Intelligent Emotions

The TP tools are inherently systemic. This is because they are effective at every level, from creating strategy right through to governing step-by-step implementation and everything in between. They are an essential asset for every leader and manager because they simultaneously shape and reinforce the patterns of systemic thought and action.

As we stated in the previous chapter, in the Decalogue approach the TP tools not only support the change process, they are the tools we use to implement actions at every level. They provide a tangible resource for leaders to reinforce the systemic nature of the work, and having such a resource is crucial to support the cognitive challenges involved. The tools are fundamental for leading a network of projects with low variation and high synchronization.

The role of the TP tools is critical in leading the transformation because they link and support all its logical phases: what to change, what to change to, and how to cause the change (Figure 11.1).

The phases of change, and the supporting TP tools that go with them, mirror the rational effort that we need to make in order to connect our *intuition* (for what needs to be changed) with our *understanding* of the new future scenario (what are we changing toward) and the awareness of the right *knowledge* we need to have in order to make that future reality happen (how to cause the change). This whole thinking pattern for change supported by the TP tools is, as we saw in the previous chapter, the following Figure 11.2.

It is not possible to trigger a better functioning of our intellectual faculties without some sort of handle on our emotions. We have come to call the ability to harness the power of the emotions to propel the rational side of ourselves *intelligent emotions*. A leader above all needs to develop these.

	Level of resistance to change	Thinking process tool
1	Disagreement about the problem	1. Conflict cloud
2	Disagreement about the direction of the solution	2. Injections
3	Fear of negative consequences generated by the solution	3. Negative implications branch
4	Lack of faith in the completeness of the solution	4. Future reality tree (FRT)
5	Too many obstacles along the path that leads to change	5. Prerequisite tree (PRT)
6	Reservations about our ability/willingness to implement the solution (and about the ability/ willingness of others)	6. Transition tree (TRT)

Figure 11.1 Thinking process tools for the phases of change.

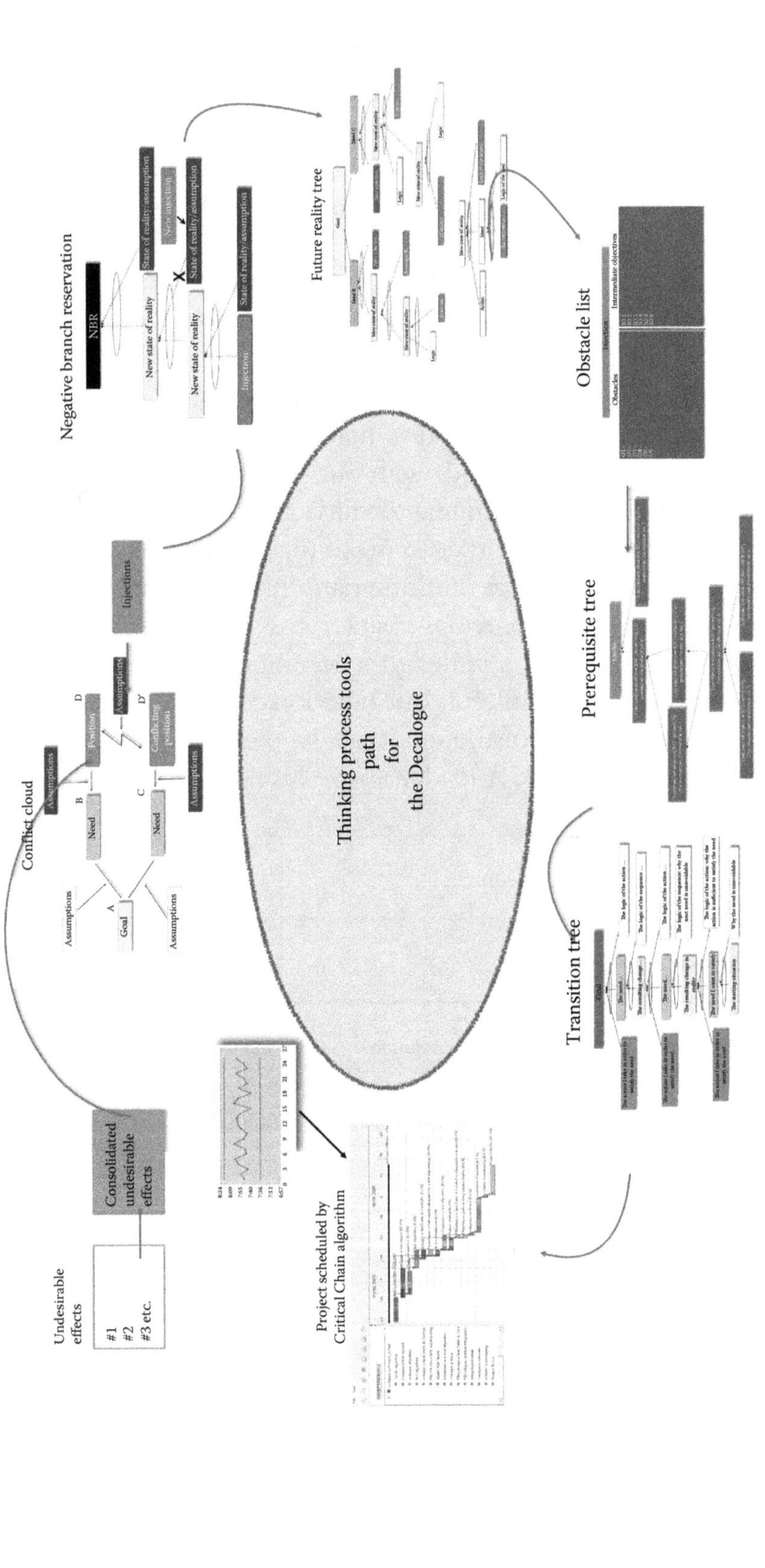

Figure 11.2 Complete cycle of thinking process tools used in the Decalogue methodology.

Emotions manifest themselves in the form of the mental models, or assumptions, that link the elements of a conflict cloud. The states of reality typified by the conflicting positions in a cloud exist as a result of some level of "amplification" of the needs for security and satisfaction; such amplification creates the "separation," that is, the conflict we experience. For instance, when we focus on our fears or decide to exercise restraint we almost invariably end up justifying "not doing" something, to maintain the situation the way it is. On the other hand, when we pay attention to our satisfaction, when we focus on the fulfillment of our desires, we become prone to wanting "to do" something. Both the neurology and the metaphysics of this phenomenon are a fascinating field of investigation.

The mental models that constitute these amplifiers, as well as the ones that underpin the conflicting positions in a cloud, are the emotions that we want to harness. Why? Because if left to themselves, they would singlehandedly jeopardize any attempt at managing the variation and the synchronization that our organization so painstakingly pursues. Unguarded and unchecked emotions can be vital for some artistic (not necessarily creative) endeavors but can easily disrupt organizations because they are powerful catalysts of entropy.

In short, leadership in a complex, network-like environment requires an enhanced kind of systemic intelligence and the mastering of the emotions that are generated by the change underway. The science of management that goes with this intelligence should adopt methods and tools that are coherent with it and the new economics is a cogent offspring of this new paradigm.

11.5 A New Economics

> Best efforts and hard work, not guided by new knowledge, they only dig deeper the pit we are in. The aim of this book is to provide new knowledge.
>
> **Deming**
> *The New Economics*

Deming's warning was clear. Unless we arm ourselves with new knowledge then no matter how hard we work we will only dig ourselves deeper into trouble. Deming provided us with this new knowledge for creating systems-based organizations and this knowledge, in turn, calls for a new economics. It is only systemic knowledge and intelligence that can enable the new economics and these need to be fostered and groomed. Before we can make it new, what do we mean by "economics"?

Economics belongs to the realm of the "social sciences." It aims at investigating the production, distribution and consumption of goods and services. Economics also concerns itself with the study of economies and how the players, the decision makers, act to guide economic choices.

At the most fundamental level, economics should pertain to the understanding of how to deal with the resources at hand and optimize their use for a stated goal. In this sense, economics is also a "political science" because the use of these resources should be guided by political decisions. (Indeed, politics should be guided by a philosophical and ethical vision, but this is another story.)

Economics mimics science by developing models, in this case economic models, that should explain the economic outcome of certain decisions and these models are, or should be, inspired by a vision of the world. In recent decades, The Royal Swedish Academy of Sciences has instituted a Nobel Prize for Economics and several economists have been awarded for the models they have developed.

In summary, an economist, hopefully inspired by a vision of the world, develops models that should guide the economies of countries to an optimal utilization of their resources for a stated goal. Governments, hopefully guided by a vision, embrace economic models based on their adherence to their vision.

Where does the problem with models lie? Any model is and must be based on a set of assumptions. When these assumptions are not verified and validated the model is bound to fail in providing the results it was designed for. Of course, the political circumstances of any democratic country change very frequently and the ability to translate models into effective policy making is always less optimal than one would wish; moreover, an increasingly interconnected world calls for increasingly complex models with assumptions that are harder and harder to validate. Indeed, governments are pressed to take actions and these actions have to accommodate for political agendas that are not necessarily driven by the vision that inspired the economic model. Moreover, when time (and perceived risk/reward) comes into the picture and we slide into the field of "finance," we witness the full potential of the prevailing economic paradigms as reflected in the models that, tragically, still today purport to create value.

If we continue the analysis of this chain of causes and effects we can understand why the world is experiencing the current economic situation.

11.6 Flawed Economic Models

We need to broaden our view of what economics should be because it affects us all, and leaders in particular. Mainstream economic and financial models, the ones that currently rule the markets and determine value, have shifted their focus over the years. Whereas previously the attention was placed on what was best for the society they were trying to model, now the emphasis is on what it is mathematically possible to achieve for the benefit of a few. It is thanks to mathematics that we understand the physical world and it is thanks to its rigorousness that we are confident that the scientific method can provide acceptable validity. Sadly, the models of most economists and financiers are far from being the offspring of any scientific method.

Current mainstream economic and financial models are flawed for two sets of reasons:

1. They are often divorced from realistic assumptions about the situation they seek to model AND from the managerial actions that should ensure the predicted outcome. In other words, the modeling happens in the vacuum of second tier "mathematical" speculations with flawed assumptions about what is possible or impossible to achieve managerially.
2. Far too many, and often prevailing, economic and financial models pursue an idea of value that is divorced from any concept of the general wealth and wellbeing of individuals and society. Such models are based on a systematically disproven "rational" behavior that is driven by the desire for individual profit. These models are rooted in the paradigm that if somebody wins somebody else has to lose. They call it "competition" and a gigantic and ineffective apparatus has been created to "ensure" fair competition.

11.7 A New Outlook on Value and Wealth

In order to re-assert economics as a useful field of investigation we have to re-ground it in a new paradigm; a new economics can only be originated by a new outlook on value and wealth. As Senator Robert F. Kennedy said more than once, "The GDP cannot be considered a measure for the standard of our lives."

The starting point is to define what the role of the government should be and which policies an economic model should mirror. Any government should first and foremost protect the freedom of its citizens, for sure: freedom from any risk of slavery. Three major factors impact our freedom, other than the ability to protect ourselves from enemies and practice the religion of our choice: freedom from ignorance, freedom from the tyranny of diseases we cannot afford to cure, and freedom to start or adhere to ventures, business or otherwise.

So, the role of the government in establishing and endorsing an economic model is clear: a solid education and research system, affordable healthcare for everyone, and a network of support for the development of any form of free enterprise.

How we build these systems, how we manage them, and what set of values should inspire them is the kernel of the new economics. Economics, then, really becomes the science that studies how countries should develop.

The new economics should not just be concerned with better mathematical models to portray scenarios; any serious mathematician would always alert decision makers to the probable fallacy of such models. The new economics must be intimately connected with the ways wealth can be created and the best ways to increase the distribution of this wealth.

The world we live in becomes exponentially more and more interconnected and wealth (and its creation) is a multifaceted entity. Which is the wealthier country, one where the GDP is high but millions of people cannot afford serious education and healthcare, or one where the GDP is lower but these "freedom rights" are guaranteed? This conflict exists only because "economics" is anchored to flawed assumptions about wealth and value.

The distribution of wealth, seen as conflicting with the right of the individual to amass personal wealth to the detriment of others, has always been labeled as "socialist" and as such unsuitable for the free world. The ugly truth is that prevailing economic and financial thinking has led to the squandering of the resources that the planet has available and to the stifling of innovation. This thinking has systematically favored short-term decisions over long-term planning. This thinking has swayed tens of thousands of talented people away from applying their minds to constructive and foundational work and toward the sterile and artificial domain of "financial products." This thinking has led us to believe that we can create something out of nothing.

Never in human history has the word "scarcity" meant so much. Our resources are scarce and we need to learn how to use them; the name of the game of any serious economic effort then becomes "sustainability." The new economics must become the science that studies the optimization of scarce resources and in order to do so must tap into the bodies of knowledge that deal with how finite resources can be successfully managed.

The new economics must also be based on the founding assumption that no win can be based on somebody losing; that we are all interdependent and the wellbeing of individuals is critical for the wellbeing of society; that wealth must be created in order for it to be distributed and any form of imbalance will soon turn into a global loss; that individual success to the detriment of others cannot be sustained. The new economics is founded on the assumption that individuals, organizations, large systems, and networks and, ultimately, countries are vehicles for the creation and distribution of ideas, products, and services that help everyone to live better, more intelligently and harmoniously with our environment.

The new economics will strive to provide not just mathematical platforms but also the practical means to achieve a meaningful life.

Chapter 12

Essential Knowledge for Leaders and Managers (So They Can Manage Systemically Without Being Experts in Everything)

12.1 Where Can We Learn the Right Stuff?

This book is about the transformation of management, from the blatant inadequacies of the organizational design based on conventional hierarchy and the functional silos that go with it toward a systemic model. Our attempt is to provide science, methods, and tools to overcome the blockages originated by this obsolete way of coping with the inherent complexity of the world we live in.

Indeed, after 20-plus years of international practice with a myriad of organizations, we are fully aware that understanding the science, embracing the methods, and using the tools for complexity requires a profound cognitive shift. Why is it so difficult?

We don't have all the answers, we don't think anyone has; but we do believe that much of the paralysis and the inaction we witness in the majority of the organizations in the western world, regardless of the monumental changes taking place at every level in our society, is because our current system of business education is lagging behind.

If we are to avoid repeating the disastrous mistakes of the past that allowed a few major players to propagate waves of ruin throughout the global economy, we need to educate our leaders and managers to think and behave differently. Unfortunately, still today business schools, MBA programmers, departments of management studies, etc. on both sides of the Atlantic keep their programs rooted in the same silo mentality that produces the current situation. Moreover, they are entangled with that mentality in a reinforcing cycle that inhibits systemic learning. They are so far failing to capture the complex, highly integrated network of knowledge needed to cope with the transformation. In addition, the universities that the business schools belong to are highly siloed in their structure. Even when attempts to offer interdisciplinary studies are made, this goes against the segmented nature of departments and teaching staff may not have a way of being rewarded for teaching students "outside" their department walls.

12.1.1 The Problem with Business Schools Today

Still today, business schools tend to offer a body of disconnected knowledge taught in a fragmented manner and within the framework of a functional understanding of the life of organizations. With few exceptions, if Deming's work is taught at all it is labeled as "quality" and this limits the understanding of his work as a *systemic philosophy* for business and economics. Similarly, while it is true that Goldratt's book *The Goal* is recommended reading in many schools, it is relegated to "operations" and students are unlikely to deduce from reading this book in isolation just how universally applicable the TOC is. (In one surreal conversation the authors were asked by the dean of a business school in Canada what Deming-and-Goldratt had to do with business.)

Even where there are some attempts to include "systems thinking" as a course or module, these lessons are provided alongside traditional financial and management accounting courses and the lingering ghost of "the invisible hand" believed to be regulating the markets. In other words, systems thinking, if offered at all, ironically becomes just another silo. This is also happening in corporations that have introduced systems thinking departments, no doubt with the best intentions but demonstrating a complete lack of understanding of how to apply systemic knowledge operationally. As long as managers are measured on local, siloed efforts they will be unable to see the real advantage of any systemic approach.

The result, as amply predicted by Dr. Goldratt in his novel *Critical Chain* set in a university, is that MBA students end up not only unable to "connect the dots," they lack the understanding of which dots need to be connected in the first place. Consequently, MBA programs and the like continue to qualify individuals perfectly capable of "raiding" a company but without the knowledge and tools to build, lead, manage, and grow one.

Let's take a look at Harvard MBA required curriculum, surely one of the most prestigious (and expensive) MBA courses. Finance, financial reporting, and control are the first topics mentioned: Finance 1 looks at "the role of finance in supporting the functional areas of a firm." So students start their journey with a view of the business world based on numbers concerning functions. It is true that managers need to know how to read a balance sheet, but the numbers of traditional finance and accounting carry little useful information about how a company is *actually* performing. (See, instead, throughput accounting in this chapter.) Next up comes leadership, including "managing the performance of individual subordinates" and "establishing productive relationships with peers and seniors over whom the manager has no formal authority." So in the lessons on leadership we immediately crash headlong into a hierarchical mindset of control. What about vision? We find the following:

"Develop a vision of the future—align the organization behind that vision—motivate people to achieve the vision." There is no suggestion here that an organization has its own emergent properties. As we have already pointed out, a vision and strategy need to be derived organically from what already exists, otherwise the only road ahead is to artificially impose something "from above" and use a carrot and stick approach to push it through. The marketing paragraph does mention "marketing decisions in the context of general management," and "the relationship of marketing to other functions" but the functions remain such. It goes on to talk about "the marketing mix—product policy, channels of distribution, communication, and pricing—to satisfy customer needs profitably." This is an approach that has been around for a very long time. From a systemic point of view, pricing can only be made when you know how the production of the good or service impacts the constraint of the organization. This is the only way to have a true understanding of profitability. Moreover, crafting the offer needs to be done in close conjunction with the customer through a meticulous process of identifying their undesirable effects and core conflict, and understanding how what we offer is a solution that eliminates some of those undesirable effects while respecting their legitimate needs. (We will look at this process in detail later in this chapter.)

The technology and operations management section refers to "cross-functional and cross-firm integration," but once again this appears to be firmly embedded in a mindset where the organization is divided up into functions. The course also offers "new flexible classrooms called 'hives'—that reshape how students think, act, and see themselves." There is no mention of what this thinking is based on and the choice of the word "hive" is interesting as it refers, subliminally, to an environment where a mass of worker bees serves a queen bee.

Granted, in term 2 there is mention of "FIELD 3 brings the entire first-year experience together by challenging students to synthesize the knowledge, skills, and tools acquired," but how are the students to do this after other elements have been presented in a fragmented way? It would seem that an attempt is being made toward some kind of integrated knowledge but there is no hint at any real foundation, philosophical or theoretical, for this integration.

It is interesting to note that neither ethics nor CSRs are included in the required curriculum. When leaders and managers acquire a systemic perspective, knowledge, and tools, then ethics and responsibility are intrinsically built into the learning process. This is because an organizational design that is systemic does not hand out privileges based on rank and does not encourage "stepping on people to get up the ladder." Subordination to the constraint of an organization keeps people's focus on the collaborative effort it takes to make things happen and compensation is commensurate with that vision. Individual performance is not something that can be assessed and rewarded because it becomes clear that no one individual can "heroically" save the day. A systemic effort is a collaborative process. Moreover, people learn to see cause-and-effect relationships and the wider implications of managerial decisions for the whole value chain, not just their own company.

12.1.2 The Business School Conflict

The challenge for academic institutions faced with the task of imparting business education is, just as for any organization, a paradigmatic one. And it is based on the same assumptions of organizational *design* and *control*. The design factor, in the case of business schools, is "the design of the learning pattern" (not just the elements of the learning pattern) and the

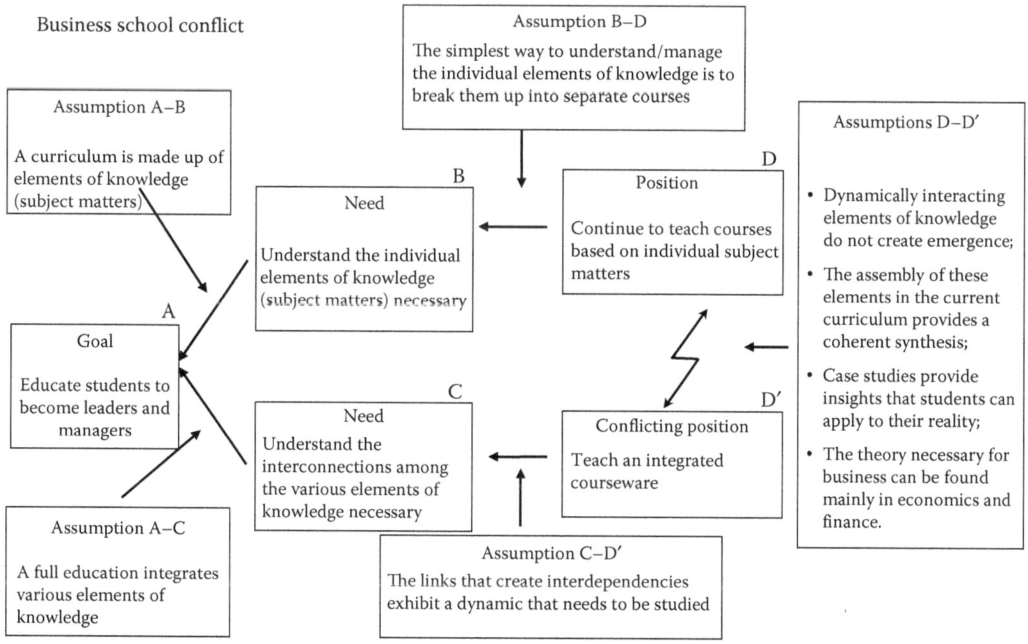

Figure 12.1 Business school conflict.

control aspect is related to the assurance that their students are hired after graduation. Moreover, business school faculties are often populated by academics who may have little direct knowledge of business as they are on an academic career path and that is a very different pattern.

Last, but not least, management is seen by business schools as something connected with economics and finance, with a smattering of operations and some marketing notions. In other words, there is little understanding of the roles that science and neuroscience play in the education of future managers and leaders. The result is the ballooning of fees for courses whose major value lies in the alumni network to which they provide their students access.

We have attempted to capture the conflict that concerns the teaching curriculum of business schools in the conflict cloud (Figure 12.1).

12.1.3 A New Systemic Learning Pattern for Complexity

We believe, just as Dr. Goldratt did, that business schools have a critical role in educating the leaders of tomorrow; but if they want to stand up to the challenge, they must develop a completely different structure and pattern of learning.

The starting point is the acknowledgment of *complexity* as the defining feature of our reality; this is the "emergent property" that has to be understood both in terms of its components and in terms of their interactions.

Leaders need to understand how to govern complexity through this blend of components and interactions; this entails journeying to the very root of what the organization they should lead is made up of and how to leverage the available profound knowledge to be leaders in the twenty-first century.

What follows is a minimally comprehensive curriculum for business schools, leaders, and managers interested in guiding the transformation of the prevailing style of management into one of whole system optimization. Although we present the elements of the curriculum below in a linear way as that is the way we all read, it makes more sense to present it first visually to demonstrate that all the elements of the curriculum are interconnected (Figure 12.2).

12.1.4 Mindset, Values, and Ethics

What are we trying to achieve when we embark on the "curriculum" or pattern of learning that we present here? Our goal, as leaders and managers of organizations should be nothing less than transformation. We therefore need to acquire the knowledge to be capable of carrying that out.

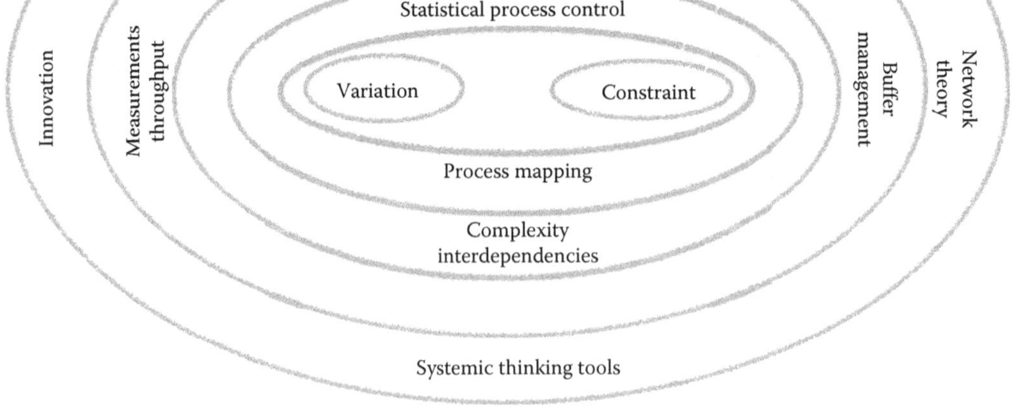

Figure 12.2 Visual rendering of the learning pattern for systemic management.

For a truly systemic organization to exist, we have to transform the way the organization pursues its goals. It is not change management, nor reengineering, nor improving, but transforming. What is required is a complete paradigm shift in the way the business of the organization is conducted. As leaders and managers, we need to learn to make that happen in organizations, the place where human effort transforms ideas and materials into saleable goods and services that bring real value. The delivery of those services and goods needs to take place as fast and as effectively as possible, providing benefits and without bringing harm to the entire value chain. This ethical stance underpins the systemic vision and practice.

Ultimately, the goal is no less than to enable a systemic attempt to provide a unified approach to economics, finance, and management, and this leads inevitably to creating the foundation for a new epistemology of wealth creation, something that through rigorous investigation separates superstition (what we believe is true and for which we have no, or a shaky, basis of explanation) from knowledge (what we know and is part of who we are).

12.1.5 A Word about Experience

For a long time, managers (usually chosen among the eldest in the organization) have been regarded as the most experienced people inside an organization. Traditionally, management ability has been considered a "quality" that a manager either does or does not have. However, this is not completely true. Age is not the right parameter to evaluate managerial ability; we can *learn how to manage* and this can be done not just as a consequence of having gained experience.

This is not to say that experience has no value and we have to avoid making mistakes we have made in the past. However, experience can only be useful when it is based on *knowledge previously acquired.*

The danger is that experience can be turned into "practice that has become theory." In other words, we generalize from experience, we induce from it: if yesterday I performed some actions in a certain way and they worked, and if today I repeat those same actions with success, I could infer that they will work again tomorrow. This kind of "linear" (and may we say primitive) thinking is not knowledge and inevitably leads to failure; this is particularly true when we understand that reality works in a nonlinear way.

As a matter of fact, there can be no successful practice at all without a robust theory behind it. Practice has to originate from knowledge that has to guide the actions we perform as the application of a solid theory.

12.2 Where Do We Start? A Systems View

We have our solid theory: it is the systems approach to managing organizations. As we stated in Chapter 1, once we have acquired a systems-based perspective, the role of leadership for complexity then becomes very clear; *to guide how the network of interdependencies occur so the organization can achieve its maximum potential.*

In order to truly interiorize these concepts so they can be transformed into managerial actions, what is the pattern of learning that we need to go through? What are the steps that guide us toward the ability to govern, lead, and manage in today's complex reality? Here, we present a path for learning to equip leaders and managers with the knowledge and tools they need to be operational and effective in a systemic way, to be able to connect the dots locally and see the big picture, and to be able to always understand the implications of their decisions within their organization and on a much broader scale. Such a program is built holistically to provide a clear line of sight on what it means to manage systemically and how to build systemic management step by step. Some elements are driven by science and algorithms, others are aimed at developing a higher sense of consciousness and awareness of the intrinsic interdependencies and connections that exist. At no point should any of the tools presented be mistaken for or approached as mere "techniques." They are the expression of a profound and precise understanding of how reality works and provide the method for interacting with that reality. Through these elements participants learn to develop their natural intuition, understanding, and knowledge of issues. It is, in essence, a curriculum for a new kind of business school for an interconnected world.

So once we have established clearly and accepted fully that we now live in a world dominated by the paradigm of systems and network; what's the first step in breaking down the walls of silo-thinking and moving toward quality, involvement, and flow?

12.2.1 Being Process Oriented

The shift starts when we realize that everything we do, at micro and macro level, in its essence is a process, with an input and an output. Boxes and walls do not naturally exist within organizations unless we put them there. Processes do. Processes of communication and delivery. Processes make up the "nervous system" that connects every element of a living organization, just the way Deming depicted it. And just as Deming pointed out, there is no system without a goal. The processes in a system are all part of an effort to achieve the goal of the system (Figure 12.3).

How can we practically capture the nature of a process? By the following:

1. Mapping it out
2. Understanding that it is part of a system and seeing the interdependencies that the process feeds and is fed by

How do we map out processes and "see" the system of an organization? We can flowchart the processes. This is a very practical and hands-on way of understanding what goes on in organization and how to improve it. Remember, however, that when it comes to hands-on matters, a leader/manager is like a medical doctor. A doctor needs to be able to prescribe and read an x-ray. They don't need to be able to take the x-ray. In the same way, a leader/manager needs to know when flowcharting is necessary and how to read flowcharting and see where quality improvements can be made. They don't have to do all the flowcharting themselves.

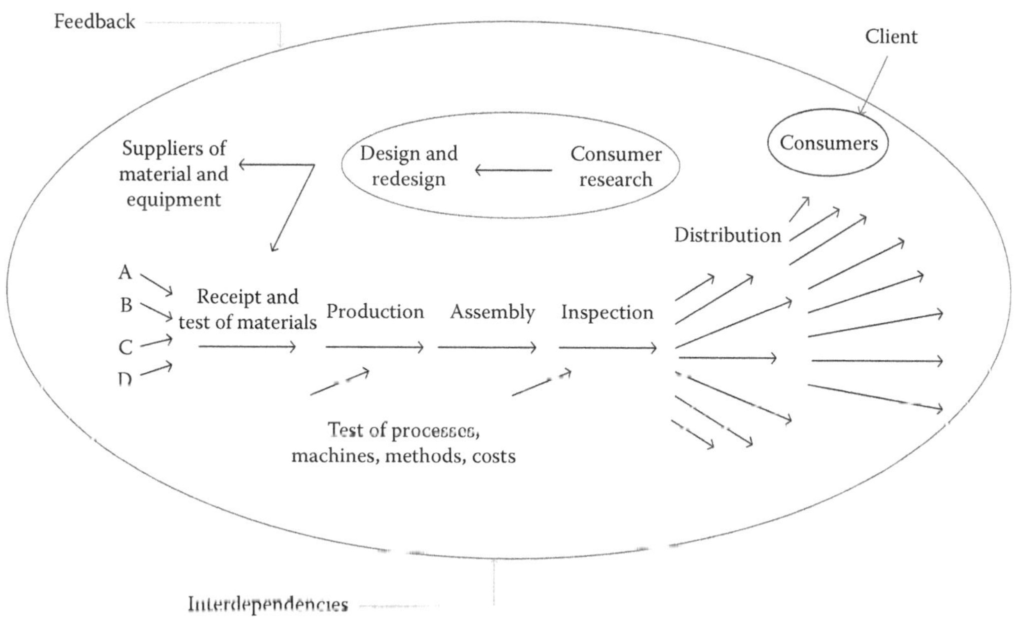

Figure 12.3 Organization as a system.

12.2.2 Mapping the System with Flowcharts

A flowchart is a diagram that describes a sequence of events, activities, tasks, and decisions that transform input in a system or process into an output. In order to draw our picture of the system, we use some standard symbols and conventions to make them easier to communicate and understand. A flowchart represents a process as a map, or as a chain of tasks and decisions.

A flowchart:

- Describes the flow of materials, information, and documentation
- Displays the various tasks contained within the process
- Shows that the tasks transform inputs into outputs
- Indicates the decisions that need to be made along the chain
- Demonstrates the essential inter-relationships and interdependencies among the process steps
- Reminds us that the strength of a chain depends upon its weakest link

From this picture, it is immediately clear that most process chains transcend the normal pyramid or "organizational chart" and are intrinsically cross-functional because the purpose is to deliver products or services to the customers.

Flowcharts are, therefore, the key to understanding where, how, and if each link adds value within the chain:

> A flow diagram also assists us to predict what components of the system will be affected, and by how much, as a result of a proposed change in one or more components.

This is what Dr. Deming emphasized in his book *The New Economics*.

It is evident that flowcharting has to be done by people that work in the processes (they are the ones that know their job).

As we flowchart a process, we think of an "ideal" way for the process to work and we immediately identify some points of improvement as a direct result of tackling:

- Complications (e.g., temporary solutions that were created to "put out fires" but that have become institutionalized)
- Misalignments between the authority and responsibility of people that prevent them from carrying out a task
- Previously hidden problem points
- Breaks in the customer–supplier chain

Most important though, flowcharts help us in identifying the "key quality characteristics" (KQC).

These are aspects of the process, which strongly affect its ability to achieve its goal. By highlighting these critical needs, we are able to identify the points where we may most valuably collect data on process variables.

From those data, we can understand the process behavior, the kind of *variation* it is affected by, and, last but not least, whether or not improvement attempts are succeeding.

Operationally, a DFC shows who does what. It shows the interactions between people along the various stages of the process. The knowledge of these interactions is vital for true understanding of how the process operates and how to improve its flow.

The improved process needs to become standard for all the people operating within it. For this to happen, the process must be documented, and the workforce must be trained in and on the improved process.

At the same time, managers must familiarize themselves with the processes in order to improve their ability to lead. In fact, flowcharting can help managers, process owners and, generally speaking, all the people working on the process, to do the following:

- Learn more about their processes by sharing knowledge and working together in order to improve them
- Work out how to do their jobs better and consistently without procedures being forced upon them "from above"
- Discover appropriate measurements/control points
- Understand the nature of variation in the process, thus avoiding tampering
- Get their processes into statistical control (the essential preliminary attempt to improve them)
- Communicate the nature and the details of the whole process to all interested parties
- Develop and deliver training materials

Basic questions:

1. What should be flow-charted?

We want to flowchart recurring processes which are keys to the achievement of the goal of the system.

Projects, for example, need not be flowcharted. We manage projects using a very specific systemic approach called Critical Chain and we will look at that later in this chapter.

2. How detailed should the flowchart be?

We want to include major interdependencies along the path of the process showing where things are exchanged (e.g., products, information, documents). We don't want to miss any exchange of material along the process.

We can produce, at any time, supporting documentation where it is necessary to have work instructions. A thorough and well-organized set of flow charts and work instructions do represent the backbone of an intelligently built quality system.

12.2.3 Getting Started

A "process flowchart" simply sets out the sequence of activities and decision points within the process. It does not show who is involved with any step. A DFC, instead, depicts who does what and the interactions between people along the various stages of the process.

That's why it is necessary to involve all of the people who work in and around the process:

- Those whvvo do the work
- The suppliers of the process
- The customers of the process
- Those responsible for the process

It is useful to designate a facilitator to do the following:

- Ask objective and searching questions to help people think sequentially
- Seek input from all involved
- Solve conflict along the flow

We start defining the process boundaries by answering the question: Where will the study start and finish?

Then we walk through the process in sequence and we gain a clear view without too much detail.

We should ask questions to prompt sequential thinking:

- What are the inputs to your task that are relevant to the process under discussion?
- Where does each input come from?
- What do you do with it?
- Where do you send your output to?
- What form does that output take?

We identify the key activities (competencies) and we list them across the top of the flowchart.

Placing the mainstream activities at the center of the page will help us to reduce oscillation back and forth across the flowchart. Otherwise, we put the activities involved earliest in the process toward the left, and those involved later toward the right.

We represent the flow moving from one competence to another with a horizontal line.

Now we can flowchart the steps of the process in sequence across the competencies, as they are performed, initially using just the rectangle (task) and diamond (decision) boxes connected by lines with direction arrows (Figure 12.4).

In more complicated structures, some components of the flowchart may operate in parallel (simultaneously), and such an occurrence will be initiated by a task box having two or more routes out (Figure 12.5).

If the process includes decision points, this will normally imply some "return-routing," causing some boxes to have more than one input.

Decisions often pose questions answerable by "Yes" or "No."

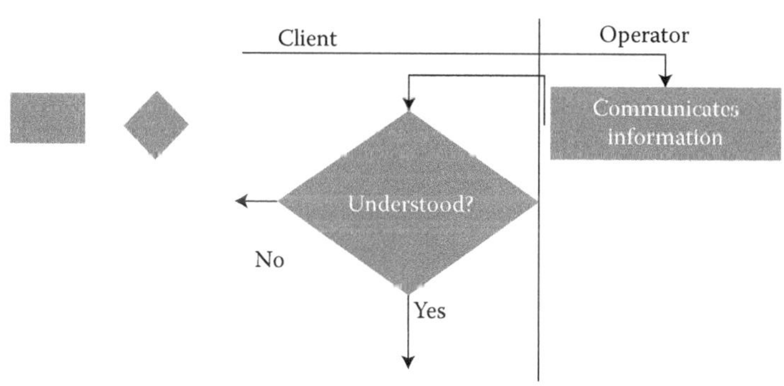

Figure 12.4 Representing flow.

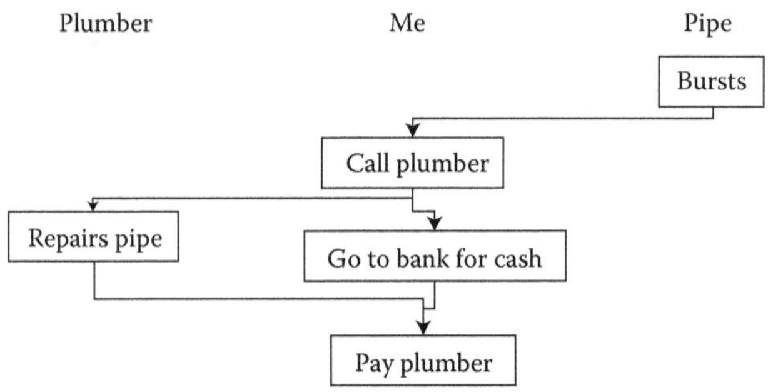

Figure 12.5 Representing parallel components in a process.

It can be convenient to structure the questions so that the preferred answering is "Yes" as "No" leads to re-routing (Figure 12.6).

A more complex DFC, relevant to a more complex process, is shown in Figure 12.7.

This is a "replenishment process." In this particular example, the "replenishment process" is the constraint of the organization.

The DFC describes all the steps, from the moment the company "X" receives the purchase order from the customer, to the moment the company "X" sends a purchase order to the supplier.

The KQCs of the process are highlighted by the small behavior charts in the picture. These charts are what allow us to monitor the replenishment of the raw material (the management of inventory) through SPC.

If we think of the organization as a network, then the KQCs provide insight into the functioning of the most important "nodes" of our network.

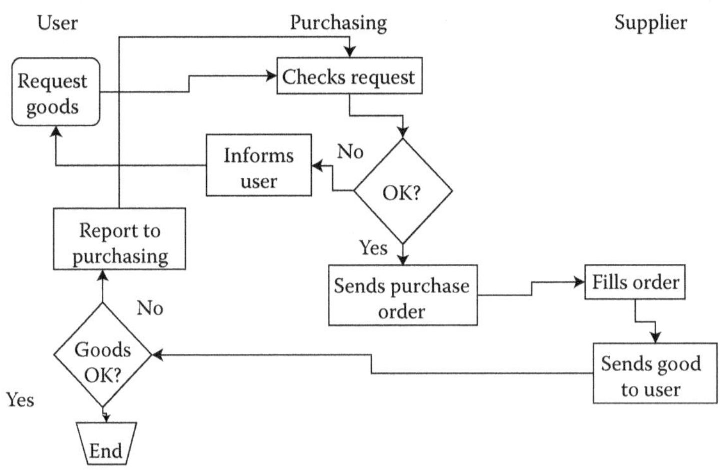

Figure 12.6 Yes and no questions.

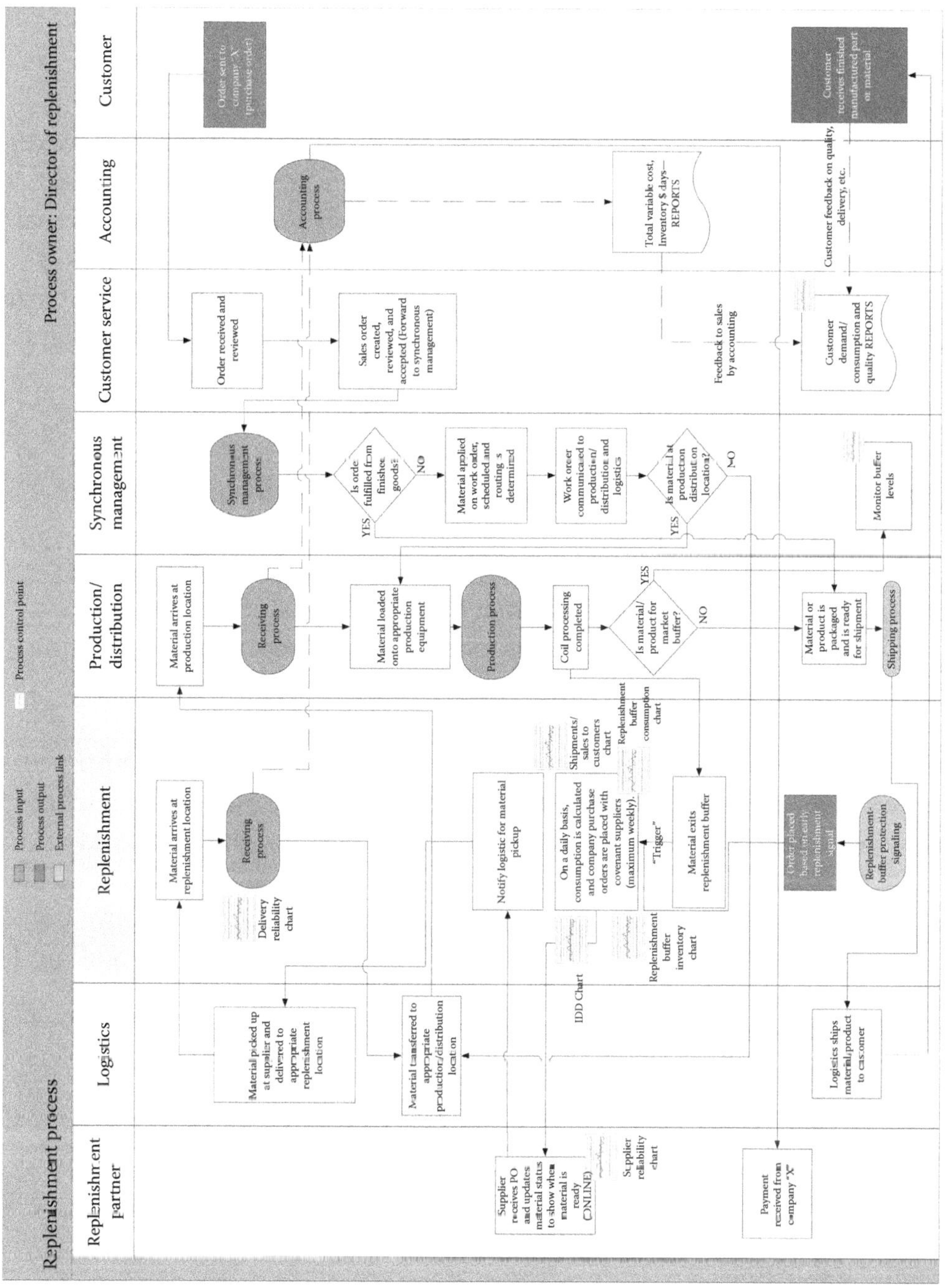

Figure 12.7 Flowchart of a complex process.

Indeed, the most important node of all is well identified: it is the *constraint* and we will talk about the constraint in more detail later.

12.2.4 Setting the Goal of the System

Understanding that our organization is made up of processes helps us make a major shift toward a systemic understanding of our organization. However, Dr. Deming stated it clearly. If there is no goal, there is no system. There is no point in mapping out a system unless we have a clearly established goal that all the processes are trying to achieve.

Choosing our goal, deciding how we want to achieve it, and communicating these decisions to everyone means essentially getting our ideas straight and aligning the values of those who work in the organization. In other words, we create a common vision. The practical problem becomes, how do we set that goal in a way that is not arbitrary or artificial? It's not enough to verbalize a goal. We need to understand what a goal is that is *suitable for our organization and its capacity*.

We can do a very good job of this by *deriving* our goal using one of the TP tools from TOC, the core conflict cloud. We looked at the process of using this tool in Chapter 10. This tool is of major importance in setting the course of an organization because it achieves several things at once. Simply using this tool accustoms leaders and managers to think and "see" the problem systemically. Then, through a thorough and painstaking process of identifying the undesirable effects of the organization and reverbalizing them as conflicting positions, the leaders can understand the profound and legitimate needs for vision and control that underpin the core conflict of the organization. The only realistic goal an organization can set for itself must respect both these needs that exist and that propel actions and mental models, whether or not the organization is consciously aware of them.

This tool is one of the most profound and useful tools a leader or manager can acquire with practice. It can be used effectively at many different levels of complexity within the organization and will always produce results that lead to informed and systemic actions that are appropriate for the specific context of the organization.

To summarize, this powerful cognitive tool will take us through the process of the following:

- Identifying our UDEs
- Identifying the true needs of our organizations (vision and fears)

- Verbalizing a realistic goal based on the above
- Thoroughly surfacing our mental models/assumptions
- Knowing what keeps us stuck and what needs to be addressed to move forward toward the goal

12.3 The Two Major Keys to Management: Variation and Constraint

Now, we are in a situation where we can "see" our organization as a system made up of processes; we know what its goal is, so how do we actually manage it in order to maximize throughput and continuously improve and innovate? In our systemic approach, this gargantuan task is greatly simplified by knowing what the two major keys to whole system management are: *variation* and *constraint*. We have to understand exactly how variation affects our organization and we have to build our organization around a strategically chosen constraint. This allows us to *synchronize* all our processes toward the goal to achieve maximum quality, involvement, and flow.

Let's start by understanding more about variation is and why it is so important to manage it.

12.3.1 Variation and Management Decisions

As we know, a system is a network of interdependent components that work together to achieve a common goal. If there is no goal, there is no system. If we have a set of segregated parts, neither interacting nor interdependent, let's call them A, B, and C, the theory of systems shows that the performance of this kind of network is "additive" (Figure 12.8):

$$\text{Perf.(system)} = \text{Perf.(A)} + \text{Perf.(B)} + \text{Perf.(C)} + \cdots$$

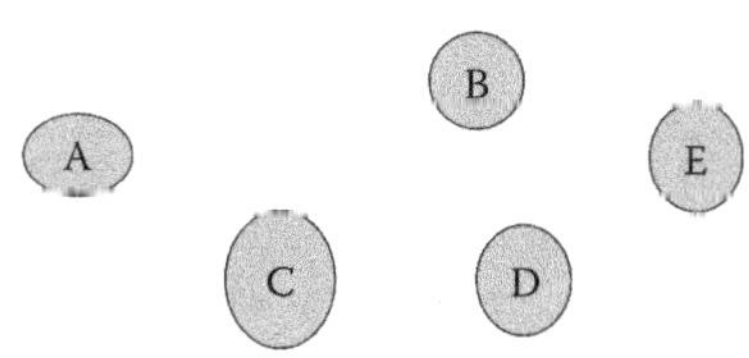

Figure 12.8 Independent network.

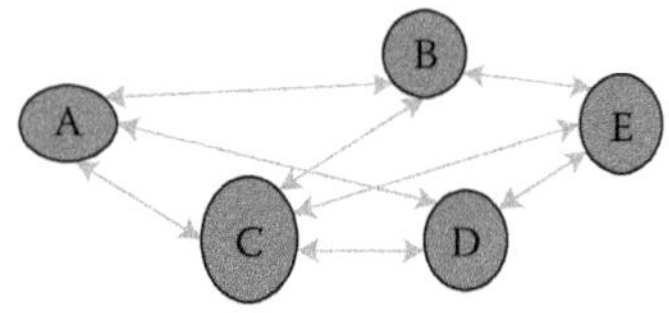

Figure 12.9 Interdependent network.

On the other hand, if the components are dependent and/or interacting, the cause–effect relations among the various parts of the system are "nonlinear," and the theory of systems shows that

$$\text{Perf.(system)} \neq \text{Perf.(A)} + \text{Perf.(B)} + \text{Perf.(C)} + \cdots$$

Interdependence and covariance are elements that influence the *dynamics* of a system (Figure 12.9). When we have large interconnected systems, it is virtually impossible to predict the effect that changing a single component will have on the system overall. We face a problem of *complexity*, where nonlinear interactions play an important role (Figure 12.10).

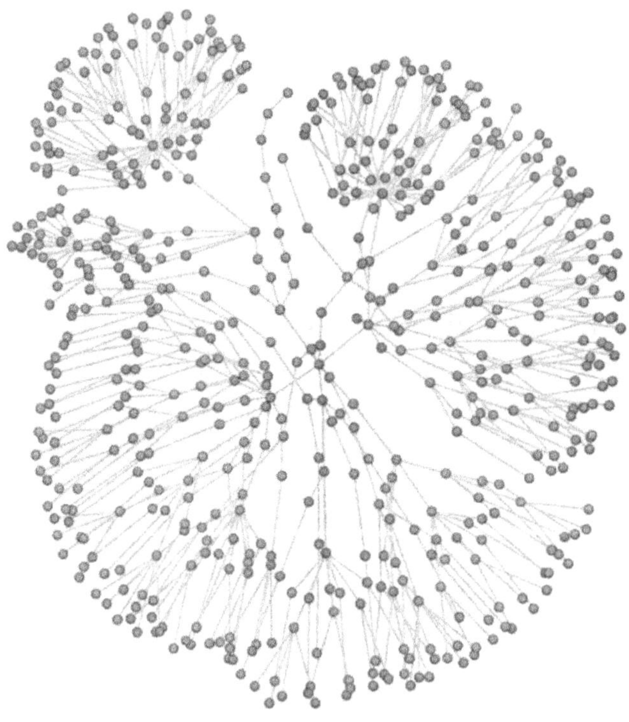

Figure 12.10 Complex system.

A company is a "complex system" ruled by nonlinear interactions that make it difficult to make any prediction when any change is effected. That's why any approach to improve the system has to be "holistic," and consider all the interactions/interdependencies.

The description of what everybody does, input and output, understanding the interdependencies and, finally, understanding variation associated with each process, is foundational to any form of intelligent management.

What is the guiding rule that should inspire any managerial decision? The job of management is to work on the system and create a predictable environment; as Dr. Deming reminds us, *the job of managers is to reduce variation.*

SPC, the main offspring of the theory of variation, is the body of knowledge that helps managers to act rationally for the improvement of the system. SPC is not a technique; it is a way of thinking that should be fostered by top management at every level in the organization. We achieve quality and the continuous improvement of the system's processes by constantly reducing the sources of variation that undermine the predictability of the processes. Predictability can be achieved by incessantly pursuing the uniformity, consistency, and reliability of our processes.

The data and the way we collect them are as important as the analysis we perform on them. The starting point to understand the system and its different processes is to select key points where we monitor variation. Once we have identified these points, we can start with the analysis. Before we look at this further, let's think about what variation actually means.

12.3.2 Understanding Variation

Any organization, or system, is naturally affected by an increase in entropy, or disorder. For this reason, the day-by-day repetition of simple actions at our work place can never be the same.

Variation is the term that describes the impossibility of repeating something in exactly the same way. It affects all aspects of our life, and all processes in an organization. It is of profound importance to managers who, in order to exert their role, must ensure a statistically stable and predictable environment. Indeed, as Dr. Deming has stated, the essence of management is prediction. Let's have a look at a few key points we have to consider when we talk about variation.

The phenomenon of variation is intrinsic to every process and system. We can define a process as a set of actions/activities that happen over time, following a rationale/procedure and aimed at a specified goal. Shopping for food at the market with our family on a Saturday is a process and so is the set of actions that get us to the office every day. Let's examine the set of actions that everybody goes through to get to work, and the relevant variations:

- How many minutes do people stay in bed after switching off the alarm? Perhaps they stayed up late the night before, so they want to stay in bed a few more minutes
- Do they make the coffee or any other family member makes it?
- Is the bathroom free or do they have to wait to get in?
- How long do they stay in the shower and how long will they take to get dressed? Did they decide the night before what they were going to wear?
- Will the car start first time or will they have to let the engine warm up?
- How many red lights and how many green lights will they come across on our way to work?
- Will they find a parking spot near the entrance to the office?

The arrival time depends on how these simple actions, with their associated variation, are combined together.

The variation associated with each action is identified mathematically by its "variance," whereas variation associated with the whole process is identified by the combination of the set of variances relevant to each action, which is the covariance. As a matter of fact, processes inside an organization are highly interconnected and interdependent, and predicting the outcome of a sequence of events becomes very difficult.

12.3.3 Variation and Processes

Let's consider the sales process of a company. "Scoring" a sale is just the last step of a process that starts with purchasing the raw material, continues through manufacturing and assembly, up to shipment. In order to be sure that we score the sale, we have to consider the variation associated with every part of the process:

- The arrival time of raw material
- Time to inspect it
- Time for moving material from the raw material warehouse to manufacturing

- Time for manufacturing
- Time for assembly
- Time for moving material to the warehouse of finished products
- Shipment time

There is no way to "predict" the final outcome of the process without considering the different steps that brought us to the finalization of the sale; the only way is to approach the process "holistically."

Understanding and improving the performance of highly interdependent processes in complex organizations is possible *only* if we understand the variation associated with single processes, the cause–effect relationships among them, and the impact that they have individually and cumulatively on the final result; in other words, we have to approach the organization (system) as a whole.

12.3.4 Working with Variation

What leaders and managers have to bear in mind is that people work *in* the system, while the job of the managers is to work *on* the system, in order to continuously improve it. Everybody is supposed to give his/her contribution in order to achieve the goal.

Specifically, what do managers have to do? In order to take the right actions and make effective decisions they have to make sure they work in a "predictable" environment. This means that all the efforts have to be directed toward the reduction of variation inside the system. But first, we have to learn how to understand the *nature* of variation the system is affected by.

There are two kinds of variation. The first kind is intrinsic; intrinsic variation is the reason why life is tough, interesting and at times comical (entropy exists). The second kind of variation, due to external causes, always takes us by surprise (e.g., an earthquake). Learning how to distinguish one kind of variation from the other is crucial for a manager. The reason for this lies in the fact that the actions we have to perform in order to deal with those two different types of variation are completely different from each other.

12.3.5 Gathering Data

In order to manage and improve processes, we need to gather relevant data and learn how to interpret the relevant data from the processes. The purpose of collecting data is to have a basis for taking appropriate actions to improve our processes.

What kind of data do we need? We start by flowcharting the processes: "If you can't describe what you are doing as a process, you don't know what you are doing," as Dr. Deming used to say. Once we have our flowcharts we can decide to apply what is known as SPC. This is lot more inspiring than it sounds. Statistics, particularly SPC, far from being a technique is a "way of thinking." This is because an intelligent use of SPC allows us to take *meaningful* actions to improve the system.

The first fundamental step is to choose what to monitor about the system and where.

Collecting data is not a simple exercise. The kinds of data, and the way we collect them, have a strong impact on our perception of how the system operates. We can choose to monitor the KQC that we mentioned in the section on flowcharting.

But given a set of data, how do we summarize and organize them so that we can "listen the voice of the process"? We have to bear in mind that our purpose is to extract useful and practical information from the data. We could use a histogram, or bar chart, that shows us the data with possible values along one axis and their frequency along the other (Figure 12.11).

But this is a very limited way to display and summarize data on a chart. A run chart, instead, shows us how the data evolve over time. Time is a key variable as far as our understanding of the system is concerned (Figure 12.12).

But we have to take a step further. In order to understand the variation that the system is affected by, we have to be able to identify the nature of

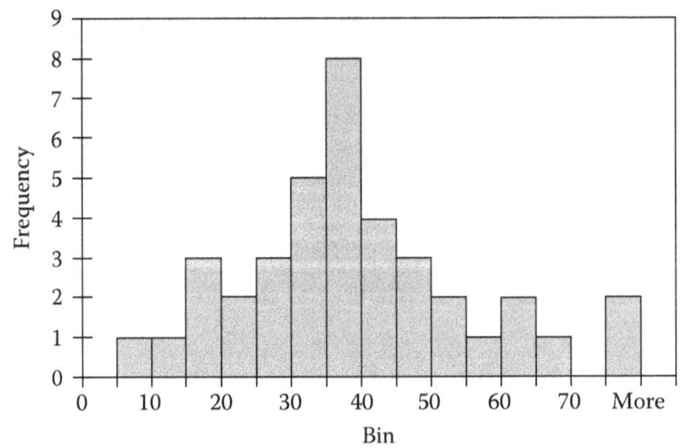

Figure 12.11 Bar chart.

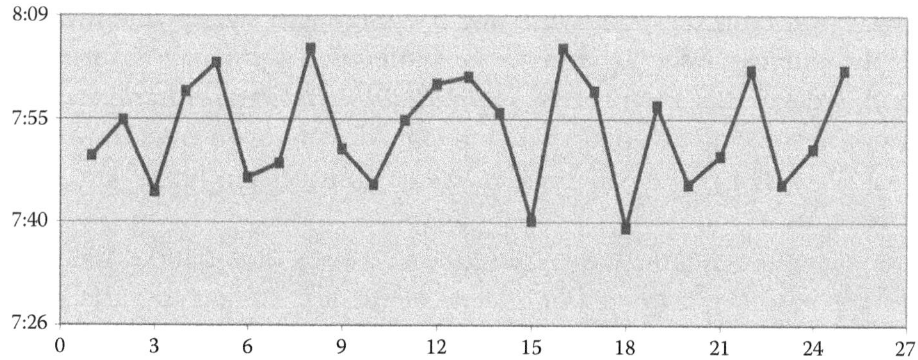

Figure 12.12 Run chart.

the oscillation of the processes, that is, to put that oscillation into a context. To do so we use what is known as a process behavior chart (also known as control chart). This shows us the evolution of the data of a process over time measured against their average value and their *natural limits of oscillation* (Figure 12.13).

So what are we trying to find out from our data in the process behavior chart? As we mentioned before, there are *two kinds* of variation. The first one is due to causes intrinsic to the process. These causes are always present, in every production batch, with every operator, in every step of the process, in every action we perform. They are called "common," or "intrinsic" causes.

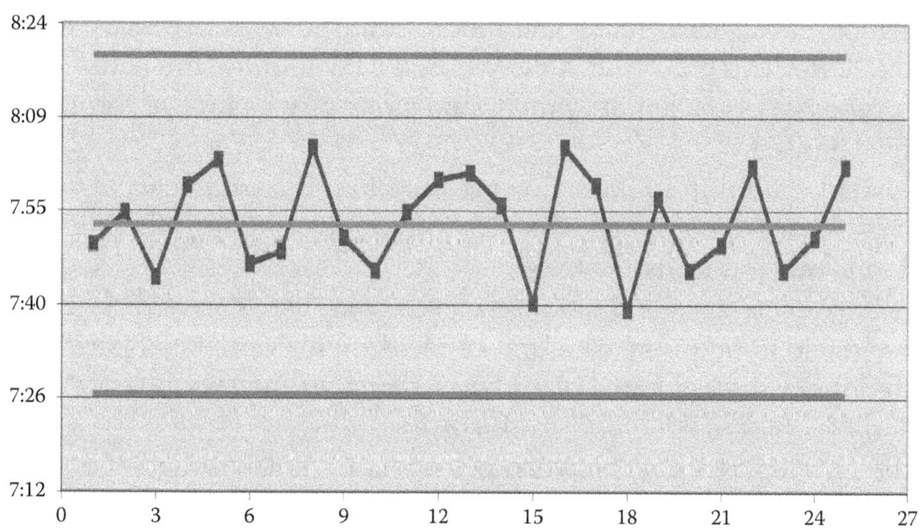

Figure 12.13 Process behavior chart.

The second kind of variation is due to assignable causes that are "external" to the process. When the process is affected by *intrinsic* causes of variation, we say that the process is *predictable (or in statistical control)* because it is subject to variation that is statistically stable over time. The behavior of such a process is *predictable* (within certain limits of oscillation) over time.

When a process is affected by assignable causes of variation, we say that the process is *unpredictable (out of control)*, as it is subject to variation that is not statistically stable over time. We cannot "predict" over time the behavior of such a process.

The word "predictability" is often misunderstood. Dr. Walter Shewhart, a physicist (and Deming's teacher) who worked at Bell Laboratories at the beginning of the twentieth century, studied the problem of the nature of variation and the behavior of processes affected by it. It is useful to remember what Shewhart said in this regard:

> A phenomenon will be said to be controlled when, through the use of past experience, we can predict, at least within limits, how the phenomenon may be expected to vary in the future. (W. A. Shewhart, *Economic Control of Quality of Manufactured Product,* Van Nostrand, New York. 1931; reprinted by the American Society for Quality Control (1980), p. 6.)

The problem was then to determine how to quantify the one and/or the other kind of variation detected, and how to make practical use of the information. It was clear that some kind of statistics was necessary to deal with the issue. The branch of statistics used to determine the nature of variation in processes is called SPC, the foundations of which were formulated by Shewhart himself.

Shewhart started from the following question: if a parameter of a process oscillates, what are the limits of oscillation for determining whether that variation is due to intrinsic causes?

The answer he gave is simple: he used *sigma*, the "standard deviation" of the variable in question, as a unit of measure for variation. Specifically, he stated that variation limits placed at 3-sigma set the "boundaries" for the variation of a process due *only* to *intrinsic* causes.

If the process oscillates between ± 3-sigma from its average, the oscillation is, in principle, due only to intrinsic causes.

Although sigma is a statistical parameter, the use of 3-sigma does not come from "statistical" considerations; the choice of 3-sigma does of course emerge from a statistical understanding of the phenomenon of variation, but it is set for strategic economical and empirical reasons.

Shewhart understood that statistics have to support the process of decision making and should not set artificial rules on how processes are supposed to behave. As a consequence, in order to make *informed* decisions and act accordingly, the question was: where do we economically decide to put the limits for the intrinsic variation of a process?

We must do this because we must have a criterion to separate the "signal" (external causes of variation) from the "noise" (intrinsic causes of variation). We don't want to mistake the one for the other and act without due reason, thus actually tampering with the system. Shewhart *empirically* decided that setting the limits for the intrinsic variation at 3-sigma was economically acceptable.

If we want to talk in statistical terms, we can say that the probability that the output of a process falls outside the 3-sigma limits "by chance" is extremely low. The knowledge we gain provides us with an operational procedure that allows us to know *when to intervene*. How do we know this? If the process oscillates within the 3-sigma limits, we say that it is affected only by intrinsic (to the process) variation and we do not need to intervene (unless we think the oscillation is so wide that it is not acceptable for the kind of activity in question). If, and when, the process shows data outside the limits, we know that something "unusual" (not intrinsic to the process) happened, and we have to check on the specific event and intervene coherently.

12.3.6 How to Make a Process Behavior Chart

The tool we use to monitor and measure the predictability of processes is the *process behavior chart*. It is worthwhile to stress again that using statistics in this way is not a straitjacket; we have to use it with *intelligence* as a "thinking" support for our actions.

A process behavior chart is drawn from the available data of the process; we collect the data, we calculate the "statistical average" of the data, and we derive from these data the inherent 3-sigma limits that allow us to build the upper and lower limits of oscillation in the chart.

Although there is no precise rule on how much data we actually need to calculate the limits, we recommend having at least 15–20 points. This does not mean we cannot draw a chart with 10 data points; we only have to understand that the limits are less reliable.

To determine if a process is affected by external causes of variation, we do not use just one criterion. A time-honored method (originally developed at GE) is the following.

Any process is affected by external causes of variation (hence we can consider it as statistically unpredictable) when

1. One data point falls outside the 3-sigma limits
2. Two data points out of three fall on the same side of the central line beyond 2-sigma
3. Four data points out of five fall on the same side of the central line beyond 1-sigma
4. At least eight consecutive points fall on the same side of the central line

12.3.7 Prediction and Forecast

The reason why any manager should go to the bother of understanding the statistical behavior of any process is because, as we have stated, *prediction is the essence of management.* In other words, in order to make rational decisions, managers should know how processes behave, now and in the future. However, knowing how processes will (most likely) behave does not fall under the category of "forecasting" but is instead "prediction."

Prediction, unlike forecasting, is a statistical concept. We cannot predict a precise outcome; *statistical predictability* means that the process oscillates "predictably" within the upper and lower limits of variation. This is the only kind of prediction we can reasonably make.

12.3.8 Stability and Quality

When a process is in a state of statistical control, we acknowledge that it is working with the minimum variation possible given the specific conditions of operation. If these conditions remain statistically stable, the process is predictable over time and is in the best situation possible.

In order to further improve this process, we can only try to reduce its variation following this procedure:

- We stratify the data dividing them into categories based on different factors, and analyze how data fall into subgroups.
- We separate the data by dividing them into various categories and treating them separately from the others.

- We can gain experience by applying the Deming continuous improvement cycle: *plan—do* the experiment—*monitor* its results (*Check*)—*learn* from the effect observed—*act.*

When a process is not in a state of statistical control, there is not much we can say about it. Indeed, its behavior is not predictable over time. It is subject to unpredictable jumps, or to uncontrolled drifts.

In order to reduce the variation of an unpredictable process, first we have to bring it back into a state of statistical control.

The steps are the following:

- Gather the data as quickly as possible to identify rapidly the special cause that generates instability in the process
- Activate an emergency solution to limit the damage
- Find out what made the special cause occur
- Implement a long-term solution

In conclusion, there are at least two excellent reasons why we do not want a process to be out of control. The first is connected with the impossibility of predicting which translates into the impossibility of planning (and managing). The second reason is connected with the costs associated with activities in a company that confuses *intrinsic variation* with *special causes of variation.*

12.3.9 Specification Limits and Behavior Limits

Ever since mankind began producing objects, we have known that there are no two objects can be precisely identical. At the end of the nineteenth century the problem was posed of how to define "tolerances" for components of products, that is, define to what extent it was possible to tolerate differences among pieces that had to be assembled together. This was the birth of the concept of "tolerances" or technical specifications, compared to which a component can only be *conforming* and therefore acceptable or *non-conforming* and therefore not acceptable.

The problem is that there is confusion between product specifications and controlled processes. The product specifications are the measurements imposed by the customer in order to declare when products or services will be acceptable to them and fit for their purposes. As we have seen, a process is said to be in control when the pattern of its variation is predictable within limits. The respect for specifications and the predictability of processes in principle have nothing in common.

Let's have a look at the possible relations between process stability and conformance to specifications. (The following categories come from Don Wheeler's book *Understanding Statistical Process Control.*)

1. The ideal state (Figure 12.14)
   A process in this state is in statistical control and produces 100% of conforming products.
   As long as the process stays in control, it will continue to produce only "good items." To achieve this state, the process must satisfy the following conditions:
   a. The process has to be intrinsically stable over time
   b. Any action we take on the process must be stable and consistent. The conditions must not be altered in an arbitrary way
   c. The natural spread of the process (the natural oscillation) has to be lower than the tolerances specified for the product
2. The threshold state (Figure 12.15)

In this case, the process is still in statistical control but produces some nonconforming items.

As the process is in control, the number of nonconforming items will remain stable over time.

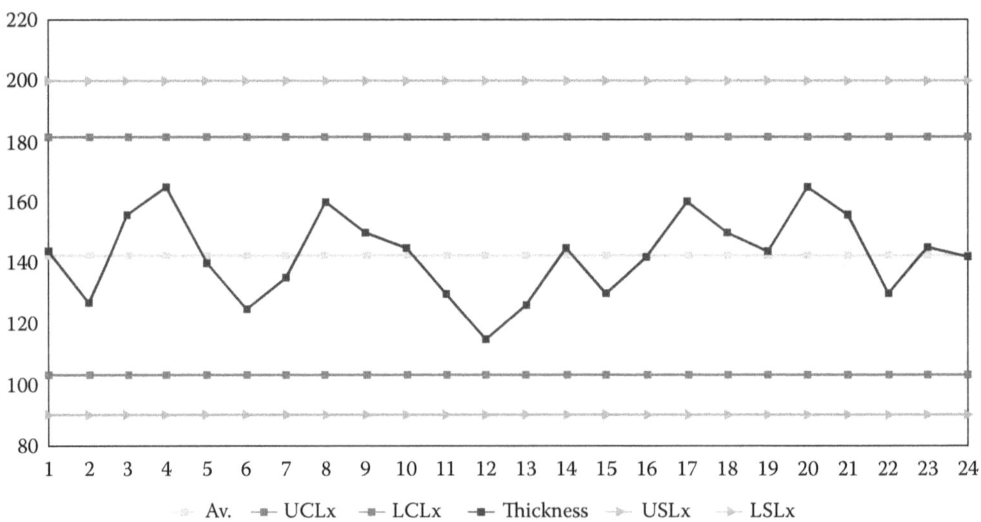

Figure 12.14 Ideal state.

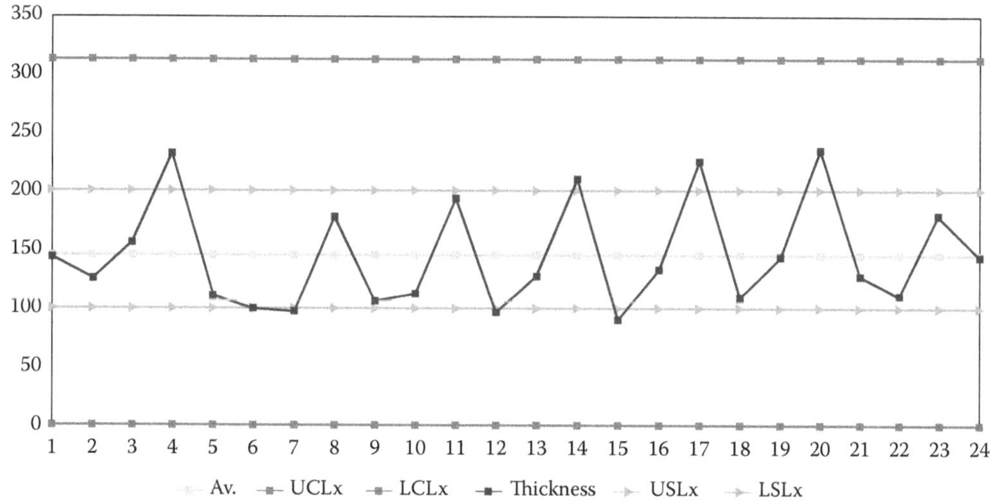

Figure 12.15 Threshold state.

There are only two ways to achieve 100% conforming products:

- Change the specifications
- Act on the process to reduce variation

In this event, we always have to remember that "tampering" with a stable process inevitably takes it out of control. This is, as usual, a "managerial" and "economical" decision to make.

3. The brink of chaos state (Figure 12.16)

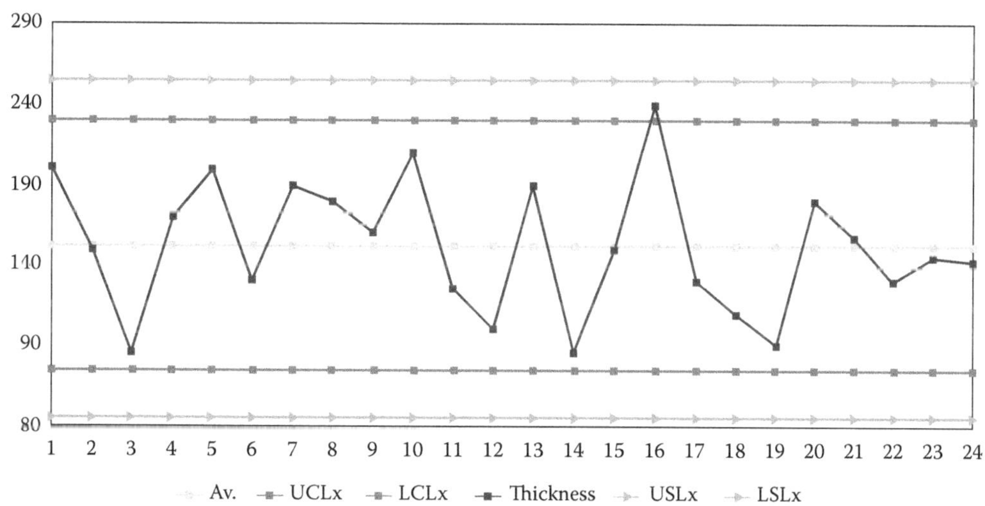

Figure 12.16 Brink of chaos state.

In this case the process is not in statistical control, even though it is producing 100% of conforming products. Everything seems to be going fine, but the process is affected by special causes of variation that make the evolution of the process unpredictable.

The behavior of such a process can degenerate at any moment, leading it to produce nonconforming products. In this event, before making any promise to our customers, we should work to stabilize the process, make it predictable, and then figure out how to comply with the specifications in a "reliable" manner.

4. Chaos (Figure 12.17)

The process is both out of control and produces nonconforming products in an unpredictable way. There is not much to say other than the only way out of this state is to remove immediately the special causes of variation and continue to monitor the process.

As can be seen, we can give the customer 100% of what he wants while operating a process, which is unstable. Conversely, we can consistently produce something out of specs while operating in a predictable manner. Which is the best? In general, we can't say. What we can say is that a state of control is not a natural state for a process and entropy *does* exist. In other words, a process that presently has an outcome that satisfies the customer, but is highly unstable, is very likely to degenerate into a chaotic

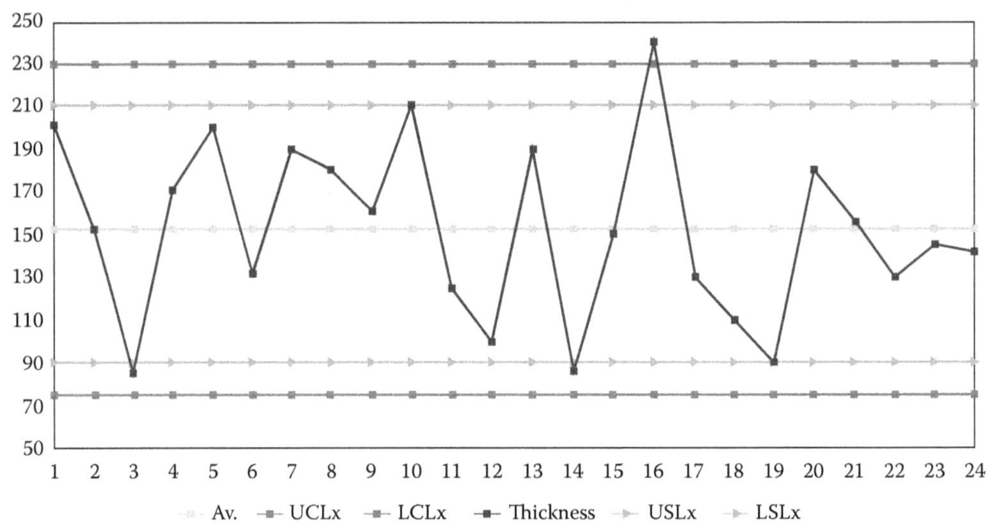

Figure 12.17 Chaos state.

state without much warning. This is the risk that we run if we operate our processes in a state, which is on the brink of chaos.

12.3.10 Variation and the Fallacies of Budgets and Performance Assessment

Reducing variation was the essence of management for Deming. It is the lack of this awareness, which accounts for much of the failure in contemporary management practice. Budgeting and budgeting techniques are examples of a management approach governed by conformance to specifications. They are based on estimates that are often arbitrary. These estimates almost invariably translate into equally arbitrary targets. The achievement of these targets, or failure to do so, becomes the basis for managerial decisions. These decisions shape the way in which people relate to the organization, the way they work and, ultimately, the way they are in the world.

Budgets, like specs in production, are "the voice of the customer." They represent what we would like to achieve, or what the customer is asking us to deliver. Neither of the two has anything to do with what we can achieve or deliver.

According to Deming, the role of a leader is to create the environment that allows people to take joy in their work, use their abilities and fulfill their potential. In order to do that the leader must eliminate fear in the workplace. As Dr. Deming points out there are two circumstances to consider:

1. The worker has achieved statistical control of his/her work.
2. The worker has not yet achieved statistical control of his/her work.

Again, the two states will call for a different set of actions, none of which entails blaming the worker. We must remember that poor performances are very often due to lack of understanding of the system that people operating in and, in many cases, the answer is in training.

Deming's work represents the most fundamental contribution to achieving quality. The goal of his work is to acquire profound knowledge of the impact that actions have on a system. This approach encourages the study of effects in order to search for the root causes which determined them. Consequently, the focus of attention opens up from the narrow realm of emphasizing results to the wider possibilities of examining processes. This is the key to achieving greater control over our actions and increasing our chances of predicting their outcome.

In Deming's approach, the ability to predict is the very essence of management. What more than anything else blocks our ability to predict? As we have seen, it is variation.

Maximum Quality is the result of minimum variation in processes. The processes which make up a system or organization are interdependent. If we do not understand the variation of processes, we cannot know what impact our efforts might have on improving a performance elsewhere. For these reasons, we need to achieve a stable system.

In his constant search for quality, Deming created a complex and rigorous theory. Applying this theory requires considerable determination and intellectual commitment on our part. In order for this extraordinary theory to yield the benefits we hope for, we need effective tools that can interpret and translate this complexity into something we can use on a practical level. TOC offers these tools.

12.4 Synchronous Management: Choosing and Managing the Constraint

We can "see" our organizations as a set of interdependent processes designed to achieve the goal of the system and we know how to monitor the variation affecting those processes through SPC so that we can make intelligent decisions about when and where to act to create stability and predictability and to make improvements.

How can we make sure that our "oriented network of processes" (our organization) can achieve its goal? In order to make all the processes work harmoniously, we need a synchronization mechanism. Why? Because the output of the process "A" is the input for process "B" whose output is the input for process "C" and so on. Without a synchronization mechanism, we run the risk of increasing chaos in the system (we detect this as an increase in lead times, work in progress, inventory and, ultimately, as an increase in costs).

A powerful synchronization mechanism is provided by TOC and this requires us to manage our organization around a *constraint*. In this kind of system everything is subordinated to the constraint phase, and therefore the constraint phase is the only one that requires detailed planning. By scheduling the constraint *we can manage the entire system*. Reducing the variation that affects every system becomes considerably simpler when the system is unbalanced around the constraint. This is because by protecting the constraint with a suitable buffer, a certain amount of variation can be absorbed, thus reducing the overall variation in the system. The constraint is the most sensitive phase when it comes to variation, which is propagated throughout the system in a nonlinear manner (covariance).

Moreover, we do not always operate in markets where the demand is repetitive and stable, so we need a certain amount of flexibility in the scheduling of production and resources in order to handle demand without delays in delivery. By focusing on one point, the constraint, we increase the flexibility of the system.

The approach of managing the constraint has been used extensively in manufacturing activities, but its application goes well beyond simple production. It is also a general approach to the management of a system with finite resources (i.e., any system in the real world).

12.4.1 The Systemic Model and Performance

By finite resources we mean people, machines, money, etc. The traditional way of managing is to extract and exploit the maximum capacity from every element of the system (to achieve maximum efficiency). According to the conventional way of managing, if we can make machines work 100% of the time, we will achieve the maximum for the system. However, this is not true for two reasons: firstly because of the *interdependencies* between the different parts of the system (interactions behave nonlinearly), and secondly because of *variation*.

As we have mentioned, for a system made of parts "A" and "B," "C," etc., the result of the interaction gives:

$$\text{Yield(system)} \neq \text{Yield(A)} + \text{Yield(B)} + \text{Yield(C)} + \text{Yield(D)} + \cdots$$

Actually, the effect of the interaction can go both ways, and the exact equation is

$$\begin{aligned}\text{Yield(system)} &= \text{Yield(A)} + \text{Yield(B)} + \text{Yield(C)} + \text{Yield(D)} + \cdots \\ &\quad + \text{Yield(AB)} + \text{Yield(AC)} + \text{Yield(AD)} + \cdots \\ &\quad + \text{Yield(ABCDE)} + \cdots\end{aligned}$$

If the parts of the system interact in a "constructive" way the system will benefit from an "added value"; if the interaction is not constructive, the system will experience a negative effect. If we take into account variation, which we know is intrinsic to any human activity, synchronization becomes an issue and even obtaining 100% efficiency from every machine becomes a fantasy.

12.4.2 Unbalancing a System around the Constraint

As we saw in Chapter 5, the word constraint may sound negative but it is very useful. This is true for two precise reasons:

1. We need a constraint to facilitate the *synchronization* of the different processes in the system (the constraint is the element that dictates the pace at which value is generated).
2. We need a constraint because an "unbalanced" system is more resilient and it is more easily managed when the impact of variation is taken into account (we saw the simulation example in Chapter 5).

The constraint is what determines the pace/speed at which a system generates units of its goal. In for-profit companies, the units of the goal are linked to the generation of cash profit (value).

A system that is unbalanced around its constraint is like a tube with one section that is narrower than the rest: there is one phase/process/group of resources with less capacity than all the others.

Why do we manage a system through its constraint? It is because an unbalanced system is simpler and cheaper to manage. In an unbalanced system everything revolves around the constraint phase, and a detailed plan is necessary for this phase only. This schedule allows us to manage the whole plant. Reducing (global) variation in an unbalanced system means concentrating on and investing in the constraint phase only, not every single part of the process.

If we have to manage a plant, increasing the productivity or improving its performance is considerably cheaper, and less wasteful in terms of time and energy, if it is unbalanced around the constraint.

Markets that are stable and repetitive over time, both regarding quantity and product mix, are not very common. That is why following the variation in demand is very difficult. By unbalancing the system around its constraint, we can achieve a *flexibility* that is much more manageable because the problem of sizing capacity (and making any changes) only concerns the constraint and not every phase of the process.

Let's repeat this crucial concept. *The constraint determines how much the system can accomplish.* Even if we choose to ignore it, this is still the case because every system has a constraint. Instead, when we consciously work with the constraint, we can leverage it to obtain the maximum throughput our system is capable of producing.

In the Decalogue approach, by leveraging the intrinsic stability of a Deming-based system we can strategically choose the constraint instead of identifying it. In other words, we can always decide which constraint is strategically more convenient to focus on and build the system accordingly. However, we can do this ONLY because we have already built a system made of low variation processes; this is why we can safely design our company around a strategically chosen constraint. Instead of cycling the five focusing steps of TOC—(1) Identify the constraint, (2) exploit the constraint, (3) subordinate to the constraint, (4) elevate the constraint, and, if the constraint has moved, (5) go back to step (1)—we can make the system grow by appropriately choosing the constraint and sizing the capacity of all the feeding/subordinating processes coherently. Again, this is possible only because the variation in the system is low.

12.4.3 Protecting and Controlling the System: Buffer Management

To recap from Chapter 8, performance of our system is ensured by its predictability but we need a mechanism to protect it and control it. This is provided by the buffer and by its management. What is a buffer? Individual processes exhibit variation; two or more together do too, this is called covariance. The effect of covariance is a cumulated variation that can result in any sort of combination of these variances. Regardless of how little individual variation the processes of the system have, we do need a mechanism to protect the most critical part of our organization, the constraint. A buffer is a quantity of "time" that we position in front of the constraint to protect it from the cumulative variation that the system generates. Simply, in synchronizing the processes that deliver the output of our organization, we ensure that what has to be worked on by the constraint gets in front of it "one buffer time ahead."

Is there a precise and unequivocal recipe to size the buffer appropriately? No, and it doesn't matter. If you understand control charts the sizing is fairly obvious; if you don't, you have bigger problems to cope with. As Dr. Deming would most probably comment: "Let's go back to serious business." The real importance of the buffer is not in the protection from disruption, certainly not an irrelevant issue, but in the possibility to exert control over the functioning of the whole organization: if our processes are all predictable, then we can truly use the buffer as a mechanism to gather succinct, effective, and comprehensive information about the "state of the synchronization" of our company. Indeed, if we monitor the buffer with the use of a control chart, then the control becomes real insight, the goal of management.

The management of the buffer entails a Copernican shift in the way we mean control because it prompts decision makers to rely on understanding and knowledge rather than hunches; it forces them into leveraging arguably esoteric statistical studies instead of salt of the earth accounting; it constrains the unbridled vibrancy of managers into the straightjacket of rational thinking. As a result of those considerations, over the decades, the repeated attempts to introduce buffer management to Top Management have failed.

12.4.4 Subordinating to the Constraint: Buffers and Variation

Subordinating means making sure the system acts in such a way that the constraint is allowed to work flat out on the right product mix. Subordinating means having an action plan for the resources, that is, a scheduling that takes into account the presence of the constraint (we must never starve the constraint). The data/elements that are fundamental for scheduling are the following:

- BOM (bill of material)
- ROUTING
- WIP (work in progress)
- Resource availability/calendars
- Buffer
- Sales orders (due date)

By definition, every minute lost on the constraint is lost forever, with an evident loss of profit. Once we have identified/chosen strategically the constraint, we create a detailed planning for it bearing in mind two things:

- The constraint must work on the right product mix.
- The constraint must work as close as possible to 100% of its capacity.

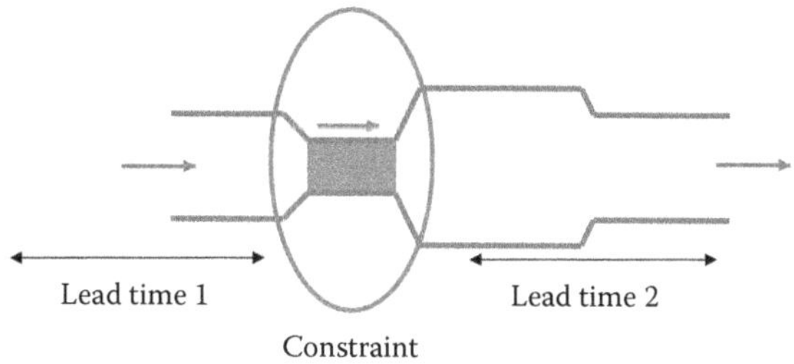

Figure 12.18 Constraint.

As far as the rest of the plant is concerned, it is enough to know, order by order, the overall lead times, that is, lead time 1 that takes into consideration the time between release of raw material and the constraint, and lead time 2 from after the constraint to the end of the process (Figure 12.18).

The orders from clients are the starting point of scheduling, which is performed working backwards from the delivery date promised to the client.

If lead time 1 = T1 min, and lead time 2 = T2 min, the DBR algorithm introduces two time intervals to protect the process that are called buffers.

Shipping buffer (before delivery date) = ~20% *T2 min Constraint buffer (before constraint) = ~20% *T1 min.

These buffers have the purpose of protecting the customer's orders by protecting the constraint from the intrinsic variation of the process. The unit of measurement of the buffer is time (Figure 12.19).

In order to protect the constraint, we have to make sure that material is ready a "buffer of time" in advance in front of the constraint itself.

In the case of buffer consumption, managers find themselves faced with a basic conflict: intervene on the system versus not intervene on the system (Figure 12.20).

The two basic assumptions, "I don't know in which cases to intervene" and "I don't know how to intervene," are injected by defining the rules of intervention (Figure 12.21).

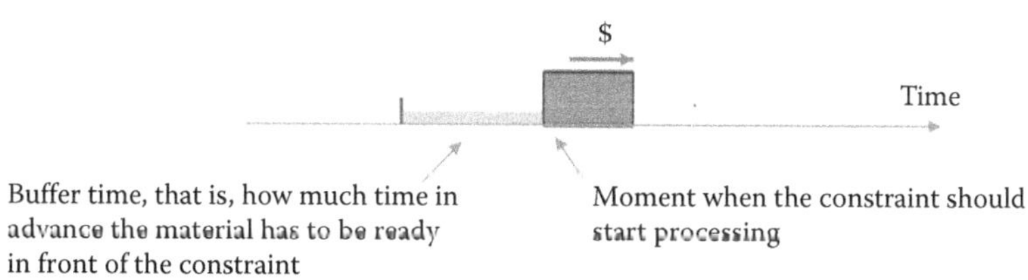

Figure 12.19 Buffer time.

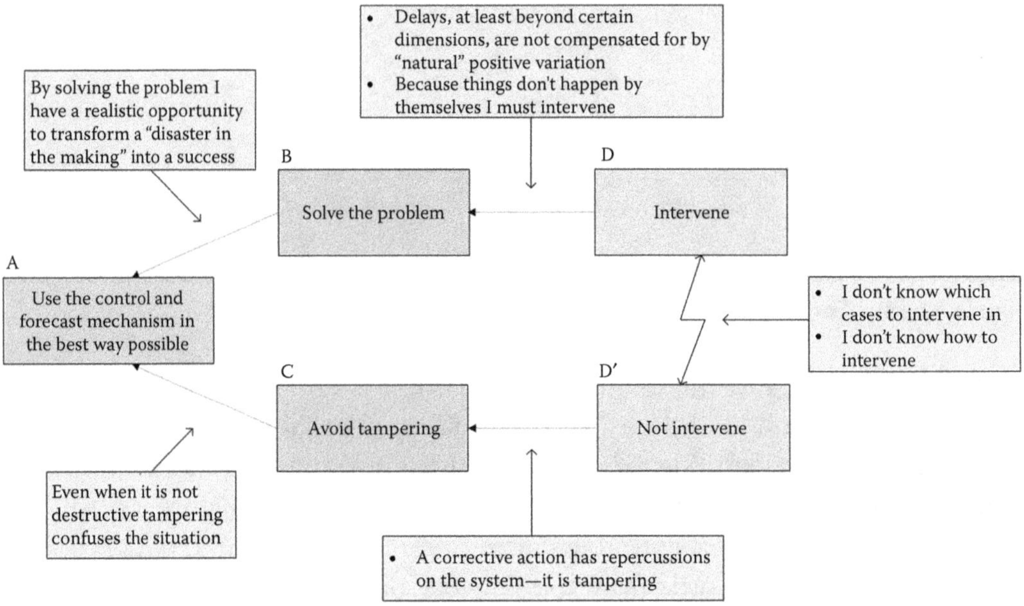

Figure 12.20 Buffer consumption conflict.

The traditional TOC approach divides the buffer into three zones and suggests different behaviors according to the zone of the buffer that has been "emptied." The buffer will "empty" out or "fill up" depending on whether tasks or operations are completed early or late in respect with the scheduling.

However, as we said, the ability to predict is the essence of management, and only if we are able to foresee the outcome of our actions on the system can we take decisions that anticipate the development of events and therefore have more control over them. In the traditional TOC approach, we "react" to a situation; when the buffer is emptied, we react. This is not the Decalogue approach.

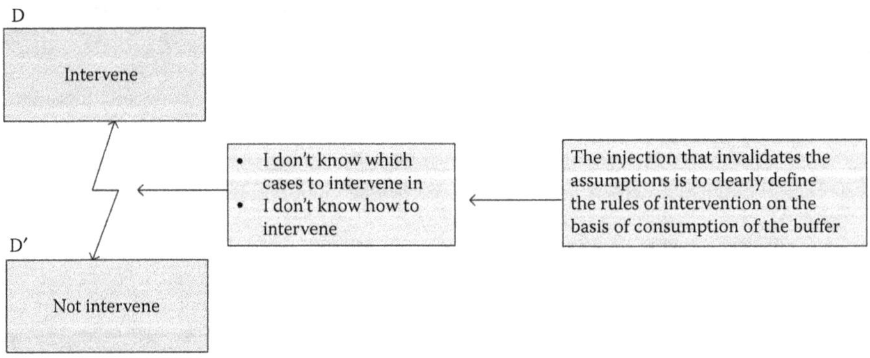

Figure 12.21 Injection to buffer consumption conflict.

12.4.5 Using Control Charts to Size the Buffer

We have defined a stable system as one that produces predictable results so, the question here is: is the system stable? In the Decalogue approach we use SPC to monitor and control the system. In our approach to buffer management SPC is used to monitor, control, and improve the way the constraint is exploited and, in general, to size it adequately.

What we do is to size the buffer, and monitor the oscillation of its consumption to see if it is in control. If this is the case, independently of how much it is emptied, we do not have to take any action. On the other hand, even in case the buffer is not emptied substantially, if the oscillation of the consumption is not in control we have to act on the system to bring it back in control because, in this case, we cannot predict what will be the consumption in the future.

In conclusion, if the processes that impact the buffer are not stable, we cannot define the amount of protection needed (sizing the buffer).

The buffer acts as a means of control *only* if the processes of the system are in control and buffer consumption oscillates stably with an upper limit that is inferior to the maximum width of the buffer (Figure 12.22).

We can summarize the "mechanics" of this process of buffer management thus:

Material is released at the pace at which the constraint can process it, so that it arrives a set amount of time early (called buffer) in front of the constraint in relation to the moment it is due to be processed by the constraint.

Order by order, a check is made so that the time when material arrives physically in front of the constraint varies in a controlled way around the established value (buffer). Variation is studied and actions are taken to reduce it (buffer management).

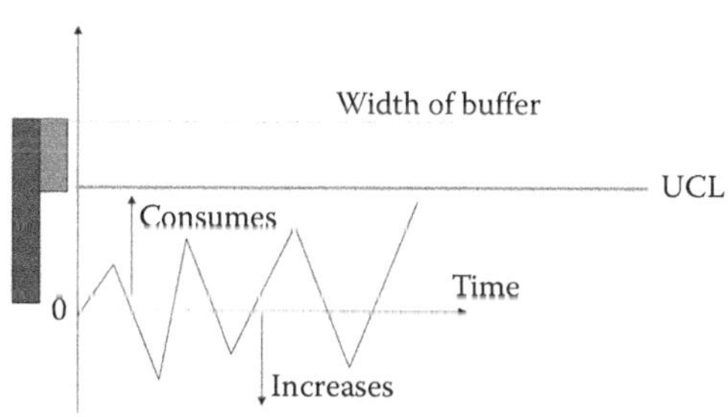

Figure 12.22 Width of buffer.

12.4.6 The Buffer and Performance Indicators

If a semi-finished piece stays in front of the constraint for more than a buffer of time, the system accumulates inventory dollar days (IDD), that is, the system holds too much inventory in relation to what it is strictly necessary to protect the constraint.

In other words, if we look at the control chart of the buffer consumption, and the width of the buffer is greater than the upper control limit, then there is more protection than necessary, which means excess IDD. However, protection is not needed only on the constraint. We also need a buffer to protect the delivery date of the product (Figure 12.23).

If the buffer that protects the delivery date is consumed completely, the delivery is delayed and the system accumulates throughput dollar days (TDD).

In conclusion, TDD is a measurement of performance related to the delivery date: if the buffer in front of the constraint is completely consumed (i.e., a semi-finished piece is not available when the constraint should process it), the deliveries are not automatically delayed, because there is still a protective buffer of finished products.

On the other hand, too much buffer of finished products becomes IDD: if this excess repeats itself in a predictable way (see the buffer control chart), then the buffer can be reduced.

Since a smaller buffer means a shorter lead-time, we can count on a commercial opportunity and the possibility to make more competitive offers to the market.

12.5 Improving Flow Company-Wide

SPC, the constraint and buffer management provide us with the focus and the practical mechanism to run our organization, but there is still a very

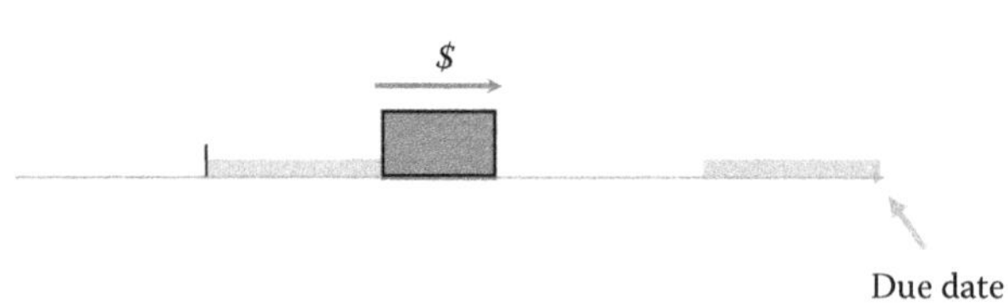

Figure 12.23 Due date buffer.

crucial element that we need because we have to be able to synchronize ALL the efforts of our organization *companywide*. We need a practical and effective way to coordinate all our resources into a flow that overcomes the artificial limitations of functions and traditional hierarchy. Our resources are not infinite, they are finite, and we have to be able to factor that in operationally to our day-to-day management. We need to be able to manage our organization as *a network of synchronized projects*.

If we want to manage our organization as a network of synchronized projects, we must have a way to avoid resource contention. Resources contribute with their competencies and these competencies are assigned to projects. What we schedule is the finite capacity (in terms of time) of the competencies we have and we then allocate the person that is available. In order to do this effectively we must have a database of competencies (name + what they are capable of doing and how well) and a finite capacity scheduler that can draw from this database only those resources that at any given time are available. Let's stress this point again: individuals with a set of competencies do not belong to a company function. They are available any time that they are not sick or on holiday and this availability is captured by the scheduler and synchronized in a way that maximizes the Throughput the company can generate. Indeed, not ALL the available time of ALL the resources needs to be scheduled, in as much as there are many activities that need to be attended to at all times by certain resources, typically shop-floor and very repetitive ones. Both levels of synchronization, scheduling of the constraint and optimization of the resources companywide, can be managed following the focusing steps provided by TOC. An example may be useful.

Let's say that we want to synchronize the work of a company producing industrial robots and automation tools. Such a company can be seen as an assembly operation and we have to decide where we want the physical constraint to be. Let's say, for the sake of simplicity, that we choose as constraint the part of the process flow where all the different components making up the final product are assembled. The focusing steps would then tell us that:

1. We release the different components at the pace at which we can physically assemble them, neither faster nor slower.
2. We ensure that ALL the components making up the customer order we want to process on that date get in to the assembly line ONE buffer time ahead.

3. Job order by job order, we monitor statistically what percentage of the buffer has been eaten into or gained. We assess the predictability of this process (that delivers the pieces to the assembly line) and then:
   - If the process is in control and the upper limit is within the buffer, we carry on.
   - If the process is in control but the upper limit is outside the buffer, we resize the buffer.
   - If the process is out of control and all the data points are within the buffer, we search for the reasons that send the process out of control and we fix them.
   - If the system is out of control and some of the data points are outside the buffer, we stop the line and fix the problem.

4. If we perform these steps and the market demand does not exceed the capacity of the chosen constraint, presumably measured in units of assembled product per time period, then we probably ship everything on time.

However, in order for this chosen constraint to maximize the Throughput of the company many of the processes making up the system (virtually all) must be synchronized:

- Engineering must issue flawless drawings
- Replenishment must deliver the subcomponents to the warehouse in time
- Accounting and administration must pay and collect promptly
- Marketing and sales must identify suitable customers and keep the line in "pull" with the highest throughput mix, etc.

There can be no "heroic" attempts here by one function to try and outshine another. Any attempt to oversell, squeeze suppliers on price, delay collection and payments, and reduce the thoroughness of drawings in the name of cost optimization will result in suboptimization of the system's performance. None of these activities can be done in isolation; any attempt by functions to outperform each other and claim more "functional" relevance in the system will not produce one more unit shipped. Accordingly, we need an algorithm and an organizational structure that helps the coordination of all the activities that maximize the throughput that we can achieve with the designated constraint.

Let's summarize. In order to manage a synchronized system we must do the following:

- Designate a physical constraint, buffer it and manage the buffer
- Orchestrate the work of the *whole* organization with a powerful algorithm—Critical Chain PM—that allows the best use of all the resources. This algorithm will be useless if we do not elect it to be the main driver of a suitable, that is, systemic, organizational structure

Generally speaking, the set of resources that is identified in the company as suitable for project-like activities should be allocated to projects for no more than 70–80% of their time. Dr. Goldratt, over 20 years ago, developed the *Critical Chain* algorithm to optimize the allocation of resources on projects and some softwares exist based on that algorithm.

12.5.1 Improving Flow through Systemic Project Management (Critical Chain)

The systemic organization approach that we present is completely founded on the concept of the organization as a network of projects. The ability to manage projects effectively therefore becomes paramount. Sadly, poor PM has to be one of the major causes of unnecessary economic losses in our world today. In spite of the existence of various official bodies and certifications for PM, the fact remains that the majority of projects fail to be completed as planned, incurring huge losses due to delays, fines, lost business, and opportunities. Lose, lose.

PM performed in a *systemic way* means taking into account not just the resources, tasks, and schedule, but how these impact each other because *resources are finite.* The Critical Chain approach to PM from TOC allows projects to be managed successfully in a way that takes into account finite resources and that intrinsically discourages the kind of behaviors that systematically undermine project completion on time and within budget (student syndrome or procrastination, bad multitasking, milestones).

Effective PM is what enables flow. Why is creating flow in an organization so important? Because not only does it provide a unique competitive edge, it also feeds the entire value chain we are part of, enabling others in it to achieve more by speeding up the way we transform ideas and materials into quality products and services. It is the opposite of hedging, that is, artificially holding back supply to the market in order to make a profit, an unethical practice for which in recent years (in)famous investment banks have been fined billions of dollars.

12.5.2 Why Projects Fail

Before we look at a systemic solution for Project Management, let's first consider why so many projects fail.

If we have to define what a project is, we have to consider that it is not actually different from a process. We can indeed define a project as a *set of actions that are required to satisfy some specifications, like those set by clients, in an established timeframe and within a precise budget.*

But why do projects fail?

The main concerns with PM are always connected with time, costs, and specifications. To be more precise, we often have difficulty in finishing projects on time, we usually have difficulty staying within the established budget and also we tend to adjust the scope and the specifications of projects to try and keep within budget and deliver on time. These elements of reality create *uncertainty in projects.*

Another important source of uncertainty is the way in which we calculate finishing dates for an activity, for a task in a project.

If we are asked to give an estimate for an activity in a project, the answer we usually give falls in a range that follows a probabilistic trend as shown in Figure 12.24.

Essentially, there is a long tail in the distribution of probabilities showing that the time estimate is affected by some strange virus. Only 50% of the answers fall between time "A" and time "B," while the remaining answers

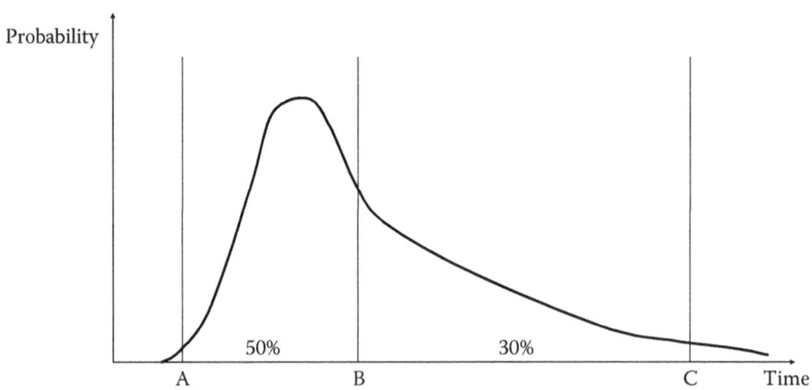

Figure 12.24 Probability curve.

are not distributed within "B" and "C" but all along the tail. Why is there such a discrepancy in the time estimate?

This discrepancy is strictly connected to the method used to measure performance.

The way we estimate the duration of a task will very often be conservative because of the following:

- We feel under pressure because our boss asks us when a particular task will be ready.
- We are (we feel) very busy (we work in multitasking mode).
- Our boss expects us to deliver what we promise.
- We feel committed to what we promise.
- We know (we feel) that something will always go wrong.

12.5.3 The Problem of Milestones and Multitasking

Indeed, two major elements that contribute to uncertainty in project completion and, consequently, contribute to project failure are multitasking and the "student syndrome."

Student syndrome is a very well-known behavior associated with the tendency to cumulate the efforts to complete a task toward its deadline (procrastination) (Figure 12.25).

It is, as its name suggests, typical in students, but it is very common inside organizations as well. Setting milestones, a common practice in PM, easily *triggers* this kind of behavior (Figure 12.26).

We think we have to work in "multitasking mode" because we know that we have to complete tasks A, B, and C as soon as possible in order

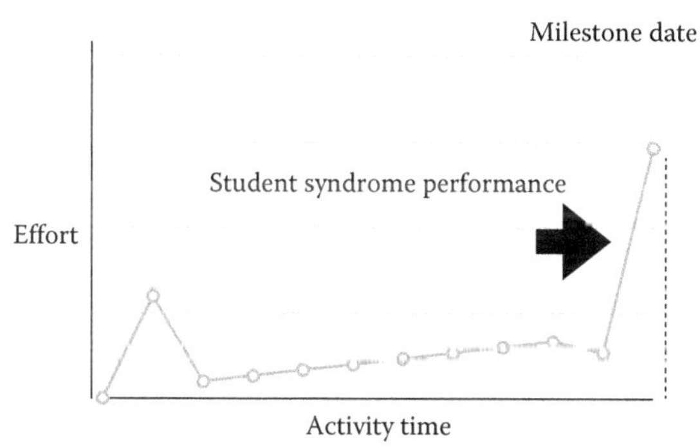

Figure 12.25 Student syndrome.

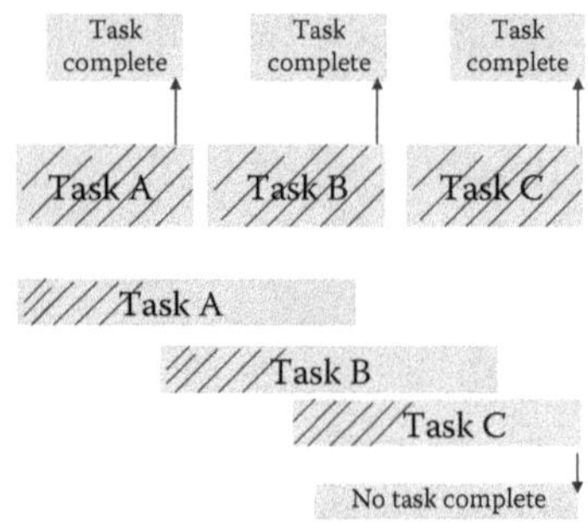

Figure 12.26 Multitasking.

to succeed. We may also receive "prodding" from the people who need to receive the output of that completed task. Consequently, we think we have to start as soon as possible working on all of the tasks to finish them and keep everybody satisfied. Unfortunately, what we don't realize is that by working in multitasking we delay the completion of *any* task when we jump from one task to another. The reality is, instead, that if we work in a dedicated way on one task at a time, we can complete all of the tasks sooner than by working partially on more than one task at the same time.

12.5.4 Interdependencies and Variation

Another feature of PM that is frequently overlooked is the impact of variation on the progress of a project.

The tasks in a project are *interdependent* and variation, whether positive or negative, impacts the project itself. Positive variation happens when a task is completed before the deadline, while negative variation happens when a task is completed later. Let's look at an example (Figure 12.27).

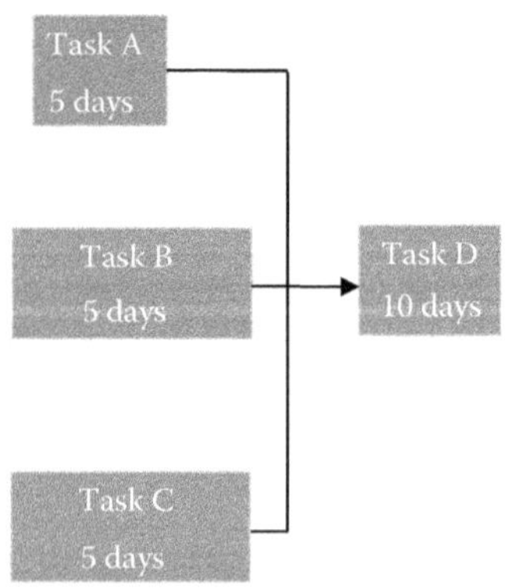

Figure 12.27 Task duration.

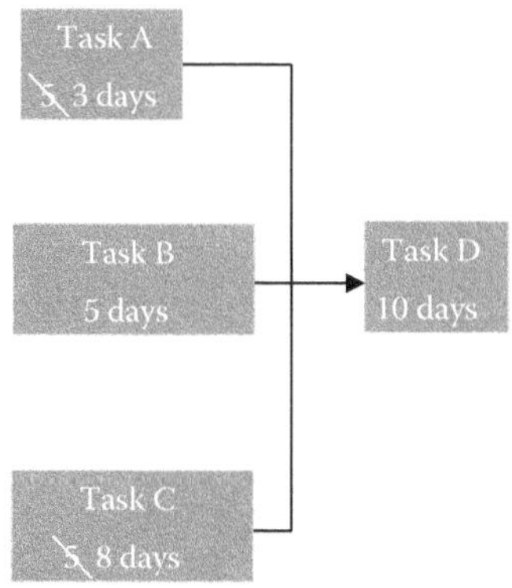

Figure 12.28 Task duration altered.

What will the impact be on the whole project if task A is completed in three days instead of five days? What will happen if task C takes eight days (Figure 12.28)?

And what will happen if tasks A, B, and C are completed in just two days? Will task D be ready to start three days earlier? We will not be able to start task D if the three tasks are not all completed. It is clear that delays get passed on while positive variation gets filtered out and lost. There is no way to take advantage of being early. Interdependencies associated with variation have an extremely strong impact on projects.

12.5.5 Measuring the Progress of the Project

A project is a network. Actually, it is an "oriented network." The goal of the project is the "cause" of its orientation. If the tasks and the resources, human and non, are not oriented toward the goal, we may continue to think that achieving a "local" success (e.g., completing a task in three days instead of five days) means something.

As a matter of fact, the completion of most of the tasks on branches of the noncritical path of a project does not translate into the project being completed on time (Figure 12.29).

The total duration of the project is determined by the critical path, not by the noncritical branches. This is the reason why placing milestones only encourages a distorted view of reality. What happens very frequently is that a customer links the achievement of a milestone with a partial payment, regardless of the real progress of the project toward the final goal. This only further reinforces wrong behavior.

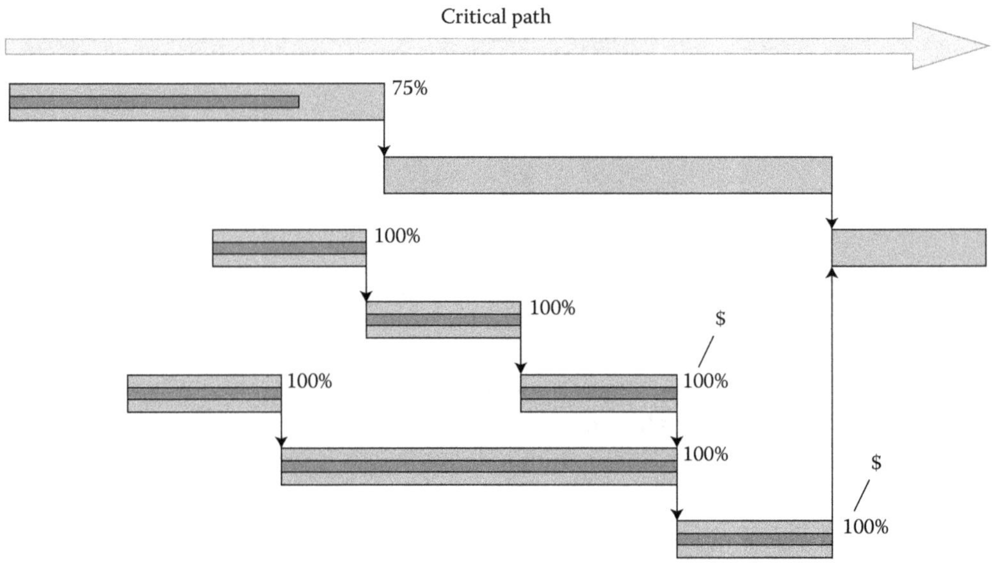

Figure 12.29 Critical path.

12.5.6 The Decalogue Approach: A Project Is a System

Let's recap on some important concepts.

A project is a *set of actions that are required to satisfy some specifications, like those set by clients, in an established timeframe and within a precise budget.* Actually, a project is a set of *interdependent tasks* that must be carried out within precise specifications.

A system is a set of interdependent components that work together toward a goal. We can definitely say that *a project is a system.*

If a project is a system, then like all other systems it has a *constraint* and it is affected by *variation.*

The Decalogue approach solves the problems of PM and therefore supports managers by providing an extremely effective set of tools that are the result of combining two powerful methodologies:

- Systemic PM using the TOC algorithm called "Critical Chain"
- Achievement of stability and predictability by reducing variation (according to Dr. Deming)

How do we proceed operationally?

We divide the project into two phases: the "planning" phase and the "rollout" phase (Figure 12.30).

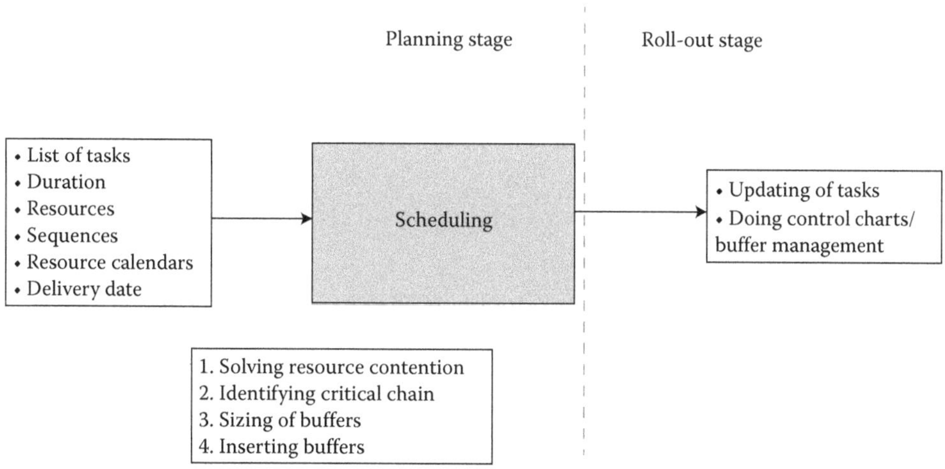

Figure 12.30 Project planning and rollout.

As far as the planning phase is concerned, the starting point is the creation of the network of tasks that make up the project. We begin listing the project tasks and highlighting the logical dependencies among them. Once the network is completed, we assign resources and a timeframe to each task, taking into account that we work with *finite capacity.*

This is a crucial point of the management of the project. Traditionally, project managers do not really take into account the availability of resources; they consider the duration of the tasks as the only parameter necessary to calculate the duration of the project. Unfortunately, we work with finite resources (both human and financial) and no one has the gift of being in two places at the same time to work on different tasks.

However, task duration is an important parameter for the identification of the length of the project and that is why we must have a criterion to assign *realistic estimates* to each of the tasks.

How can we do that? We estimate the duration of every single task by taking into account that the resources assigned to it work in a *dedicated way*, with no multitasking allowed. We assess when the resources are available and we define a delivery date for the completion of the project.

After filling in the network with all these inputs, we are going to find ourselves, inevitably, faced with resource contention for resources that may not be available on some dates or that are assigned to different tasks at the same time. The last step to complete the planning phase is to solve the resource contention and obtain what in TOC we call the *Critical Chain.*

We give the name Critical Chain to *the longest sequence of dependent events (tasks) taking into consideration the sharing of resources.* This sequence determines the length of the project. This is the "limiting factor," in other words, the *constraint* of the project.

As we have underlined many times, the constraint is the key element of the organization that has to work effectively all the time and that therefore must be protected the most.

How do we do that in PM? With the "project buffer" that provides us with the necessary protection to manage the entire project successfully.

We don't need to protect every single task, *we only need to protect the sequence of tasks that determine the length of the project (the Critical Chain).* In other words, we cumulate the protection at the end of the critical chain in *one* buffer, the project buffer, because the critical chain is the constraint that we have to protect.

We also avoid buffering every single task for statistical reasons. We know that to protect the whole chain we need a buffer that is much smaller than the sum of the buffers we should use to protect every single task. The size of the project buffer goes with the inverse square root of the number of tasks.

Finally, we have to size the buffer. Deciding the size of the buffer is, usually, a managerial decision. It has been proven effective to size it as a percentage of the total length of the critical chain, between 30% and 50%. But more than its size, what is important is how we monitor its consumption. And this activity pertains to the second phase of PM (Figure 12.31).

We are ready now to go into production (Figure 12.32).

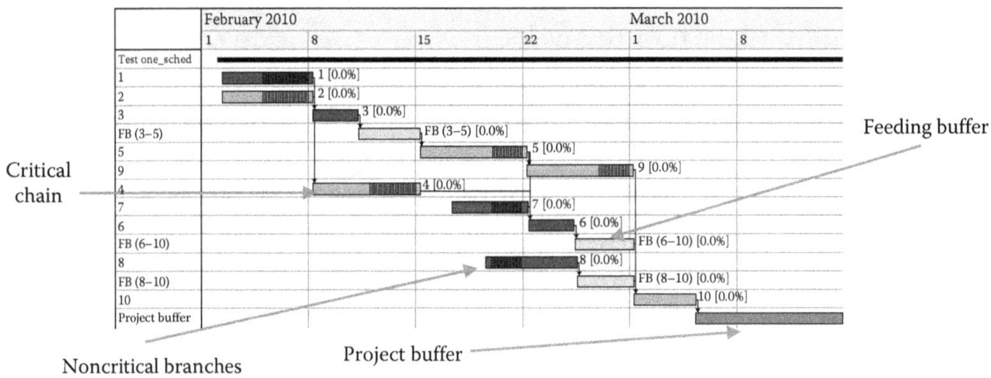

Figure 12.31 Critical chain.

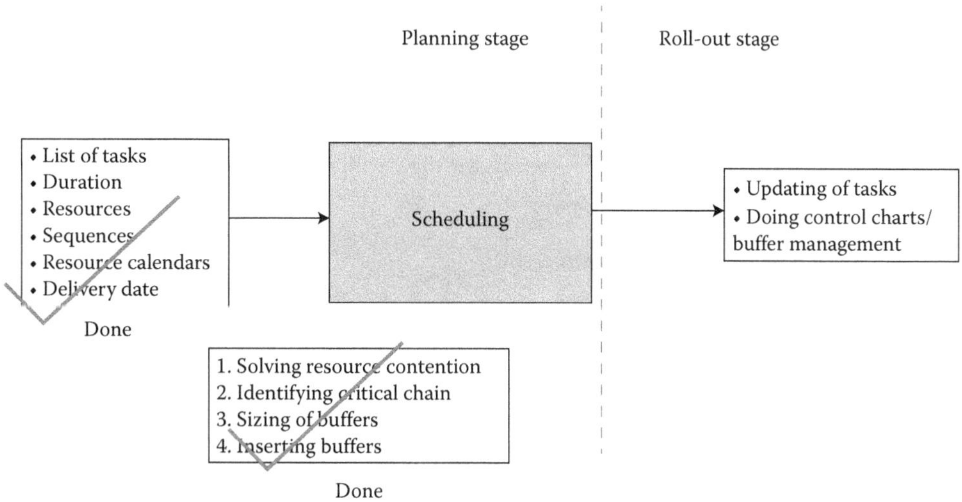

Figure 12.32 Project ready to go.

It is useful to remember that management is an extremely *demanding exercise of understanding and managing variation inside an organization.* This is also true, of course, for a project.

The impact of variation on a project is detected by monitoring how we consume the buffer (Figure 12.33).

The purpose is to have a "statistically stable system," because this is the only way we can make predictions on the completion date of our project. If the processes that influence the buffer are not stable, we cannot even know if the amount of protection (the buffer) we put in place is going to be sufficient (Figure 12.34).

The buffer works as a monitoring mechanism only if the processes of the system (in the case of a project these are the tasks) are in control

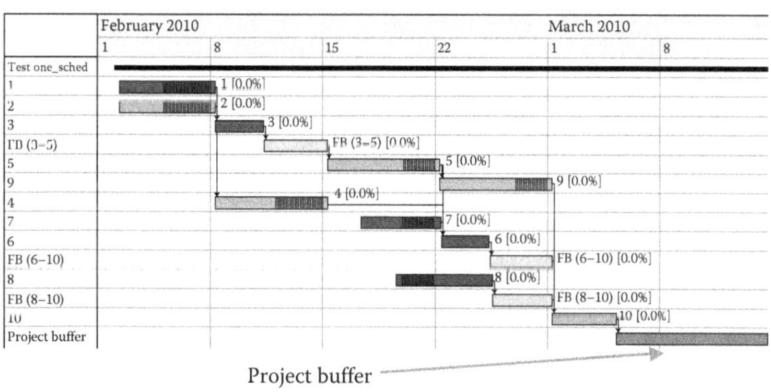

Figure 12.33 Project buffer.

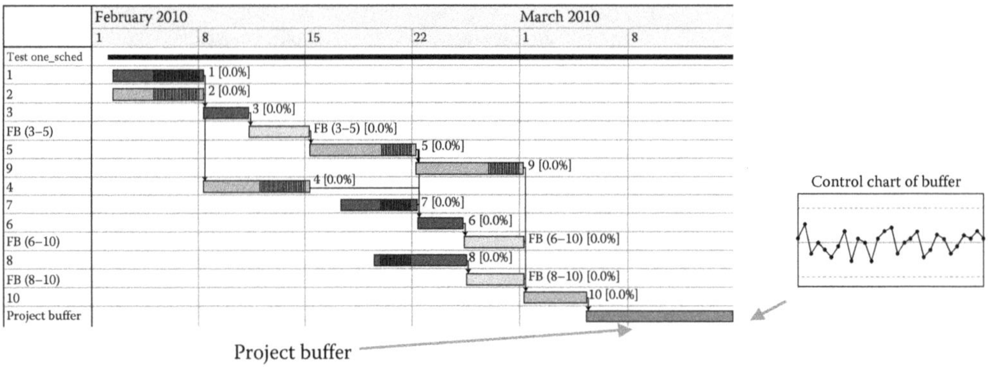

Figure 12.34 SPC and the project buffer.

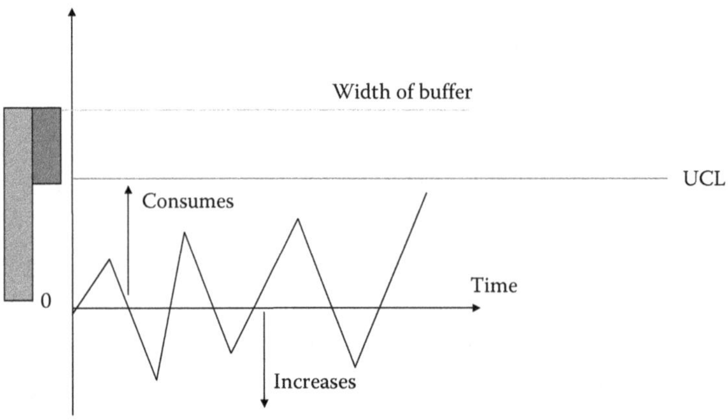

Figure 12.35 Measuring buffer consumption.

and the consumption of the buffer oscillates in a (statistically) stable way with an upper limit that is inferior to the maximum width of the buffer (Figure 12.35).

As always, the challenge of managing projects effectively is not just a question of adopting better techniques. It is a matter of correcting wrong behaviors and adopting a systemic mindset and a vision of the project as an *interconnected whole*, not as something made up of separate parts.

12.6 Measuring Our Progress Toward the Goal the TOC Way

Our organization has a precise goal. How do we know that the organization is doing what it needs to do in order to achieve that goal? It is extremely

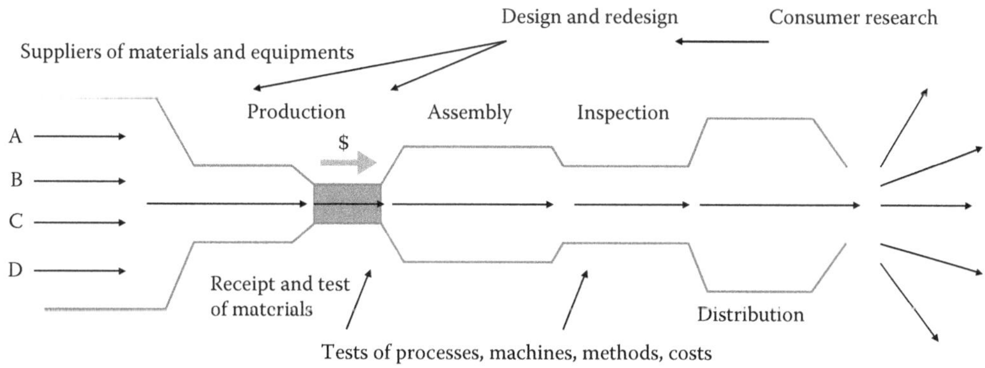

Figure 12.36 System built around the constraint.

important, in order to lead the organization toward the goal, to have clearly defined both the goal of the system and the operational measurements associated to it.

We know that variation behaves like a virus because it spreads through processes and people, combining in a nonlinear way (covariance dictates the result of the overall oscillation of the processes). As a consequence, the only way to define effective operational measurements is to consider *global* as opposed to local indicators.

The ability of the organization to achieve its goal is limited by a very small number of factors, indeed we can say it is limited by only one: *the constraint* (Figure 12.36).

The constraint determines the pace at which the system generates units of the goal, that is, *the throughput of the system.*

In for-profit companies, the units of its goal are linked to the generation of monetary profit (Figure 12.37).

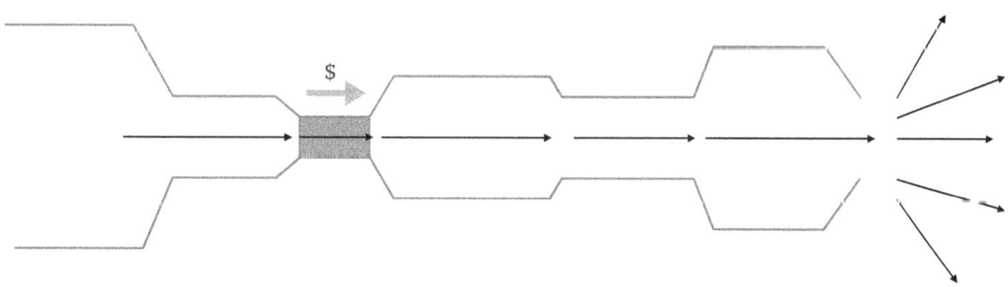

Figure 12.37 Constraint and throughput.

12.6.1 Constraints, Dynamics, and Organizations

The first chapter of any book on analytical mechanics always starts with the definition of "constraint." The line of reasoning follows the idea that any dynamical system is *constrained.*

For example, a particle moving on a surface is "constrained" by the surface to move only along certain directions; the surface "limits" the degree of freedom of the particle (Figure 12.38).

This limitation, though, doesn't mean that the particle cannot move. As a matter of fact, we know that motion is always possible and, once we take into account the actual forces, the particle follows a path that is known as *the path of least resistance.*

Organizations are dynamical systems. We can always identify (choose strategically) the constraint of the organization in order to manage it, that is, to find the *path of least resistance* for it.

This is the essence of TOC.

Anchoring the system to one point, the constraint, and understanding interdependencies (the actual forces in a dynamical system) is the way that we manage and lead the organization toward maximizing throughput (value), and this corresponds, in dynamics, to the path of least resistance.

12.6.2 Indicators from the Theory of Constraints

The logic we use in our approach is always the same: to strategically direct, support, and monitor the decision-making process connected with the

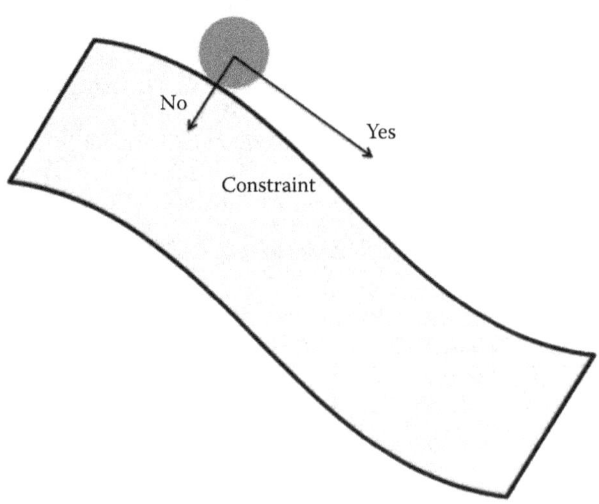

Figure 12.38 Movement of a particle.

generation of throughput. This is precisely the aim of throughput accounting. By adopting throughput accounting, we gain a much more accurate, and *systemic* picture of what is really happening in our company in terms of money in and money out. It also provides us with a much more accurate idea of the real profitability of any goods or service that we sell because it relates it the constraint of the organization.

Of course, when we adopt throughput accounting we cannot abandon tax, legal, financial, and general regulations and requirements that are mandatory in any country. On the contrary, we will have to use them alongside throughput accounting measurements.

The basic indicators we use in TOC are the following:

Totally variable costs (TVC): the monetary value of the raw material that goes into the product that we sell;

Throughput (T-put) (T): is the pace at which a company generates cash through sales. This is not a "static" measurement, in other words, it is not just the amount it generates but also the speed at which the amount is generated.

Inventory (I): it represents all the raw material and services purchased that are necessary for producing the products and services that the company intends to sell, that is, *Sales (S)*;

Operating expenses (OE): they represent all the money the company spends to transform goods and services purchased into sales. They are essentially made of two components: fixed costs and investments.

Fixed costs are the expenses that the company sustains regardless of economic results and short-term production volumes. There is little or no room for their reduction (think, for example, of salaries, leasing payments, mortgages, cleaning, and ordinary maintenance).

Investments are costs that the company plans and sustains in order to achieve long-medium term results (i.e., extraordinary maintenance, purchase of machinery and equipment, consultancy expenses, training of staff, marketing campaigns).

There are two basic equations in throughput accounting:

$$T = S - TVC$$

We can envisage this as a flow (Figure 12.39)

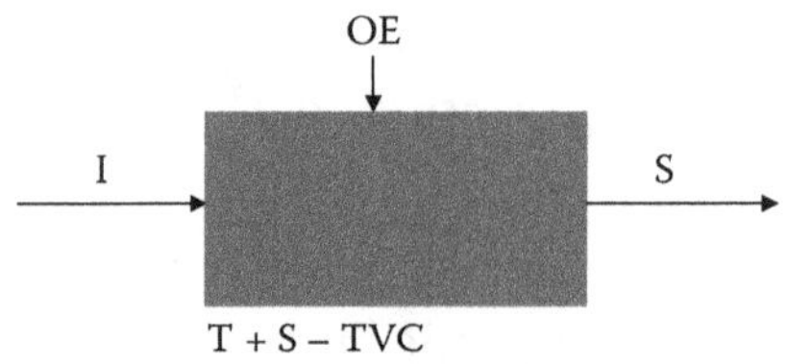

Figure 12.39 Throughput and flow.

and

$$\text{Net Profit} = \text{T} - \text{OE}$$

The difference between throughput and OE reflects the amount of money left in our hands after deducting the money we pay for OE from the money generated by sales. We must ignore the traditional measurements of cost accounting and consider the difference between T and OE as a close and accurate approximation of net profit.

If the difference between throughput and OE is positive, then the impact of the net profit on the system will be positive, irrespective of the type of financial reporting the system is subject to. This is an important point as, in many cases, managers have difficulty with the connection between their actions and the bottom line results of the company. Having the simple, practical ability to connect actions to results may upset the financial experts; they may argue that the simplistic definition that TOC gives of net profit is mistaken. The financial definition of the net profit of a system may be more complex, but any system can be considered healthy if throughput exceeds OE.

12.6.3 A Brief Digression on Speed and Time

Why is it so important to underline the time-dependence of throughput? The fact is that we live in a dynamic reality. Nothing we do stays the same. It changes over time and *when* we make decisions and take actions is of crucial importance. It can mean the difference between staying in business or not.

Let's expand a bit on this.

Throughput is a "function" of many variables. We can say

$$\text{T-put} = \text{Sales} - \text{TVC} = f(t, \text{TVC}, \text{Mkt}, \ldots)$$

$$\text{Speed} = \frac{\delta \text{T-put}}{\delta t}$$

Figure 12.40 Formula for the time derivative of throughput.

What is important is the speed at which we generate money (T-put), that is, its time derivative (Figure 12.40).

We can view the derivative as a ratio (Figure 12.41).

Finally, we can see all this in Figure 12.42.

It is evident from the graph that generating the same T-put (actually ΔT-put) in a shorter time will allow us to cover OE in a shorter time, thus freeing up cash to be used for investments.

12.6.4 The Constraint and Cash Generation

We now have a very clear definition of what throughput is: it is the *pace* at which the company generates units of its goal (cash) through sales.

$$\frac{\Delta \text{T-put}}{\Delta t} = \frac{\text{T-put}\,(t_2) - \text{T-put}\,(t_1)}{t_2 - t_2}$$

Figure 12.41 Time derivative of throughput as a ratio.

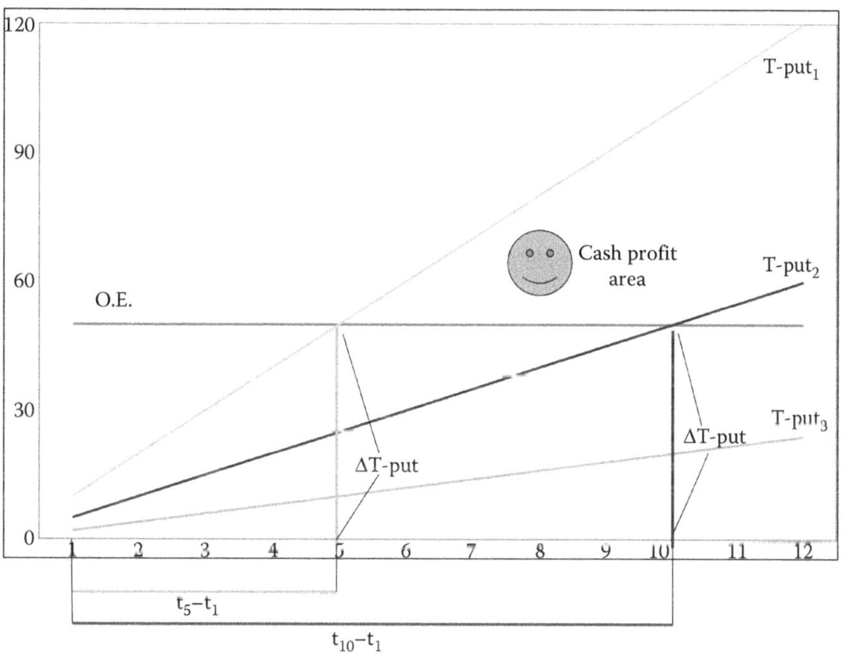

Figure 12.42 Speed and throughput.

Actually, this pace is determined by the limiting factor of the system, the constraint, and the way the constraint itself is managed. It is therefore evident that every minute of "constraint time" that gets lost corresponds to an amount of cash that we are unable to generate and that we can never regain.

If the constraint determines the pace at which we generate cash, this therefore implies that the profitability of a product is closely connected with the consumption of constraint time.

We may find our organization being faced with two different situations:

- The organization is unable to sell everything that it produces or could produce; we identify this situation as a *market* or *external constraint* (we will deal with this issue in detail further on).
- The organization is unable to satisfy market demand; we identify this situation as an *internal* or *physical* constraint.

Let's look at some practical examples.

12.6.5 Situation 1: Market Constraint

This means the market demand is lower than the products we are able to produce.

The right choice is always the one that allows us to generate the most profit for the company:

We produce a product A and a product B as shown in Figure 12.43.

In this case, the correct action is to privilege and promote the product(s) that generate the highest throughput.

12.6.6 Situation 2: Internal Constraint

This means the organization cannot satisfy market demand.

	Product A	Product B
Sales	100 $	90 $
TVC	50 $	60 $
T-PUT	50 $	30 $

Figure 12.43 Market constraint table.

	Product A	Product B
Sales	100 $	90 $
TVC	50 $	60 $
Constraint time	10 mins	2 mins
T-PUT/ constraint time	5	15

Figure 12.44 Internal constraint table.

In this situation, we have to consider the throughput generated by the product in respect with the constraint time the product absorbs (Figure 12.44).

In this case, the most profitable product(s) will be, therefore, those with the highest T-put/time ratio.

As a final example of how to use TOC measurements correctly, let's go through the first step of an exercise known as the P&Q exercise, first shown by Dr. Goldratt in the early 1980s to prove the fallacies of any form of accounting that does not take into consideration the idea of the constraint. There are many variations of this exercise and any organization could easily tailor it to its needs of understanding real product profitability. The example below was first described by Dr. Goldratt in his seminal book, *The Haystack Syndrome*. It never ceases to amaze us how little inroad this disconcertingly simple yet extremely powerful exercise has made in boardrooms and C-suites all over the world.

12.6.7 The P&Q Exercise

We are in an ideal world, and we have to make the decision to buy or not to buy a new production plant, based on its profitability. In other words, we want to understand if this plant is capable of generating revenues. The general conditions under which we have to make our decision are the following:

- Market demand is fixed.
- The moment we finish producing a product we sell it and get money.
- The moment we buy some raw material we receive it, and we pay for it.
- We work eight hours per day, five days per week.
- We have fixed and 100% reliable production time, no strikes, no machines break down, zero machine set-up time, and no overtime allowed.

Essentially, we have eliminated variation from the equation.

The plant can count on three machines, machine A, machine B, and machine C.

The production flow is represented in Figure 12.45.

The plant can produce two products, P and Q. The market demand is fixed, 100 units per week for P and 50 units per week for Q. The selling price is 90$/U for P and 100$/U for Q. We need three different raw materials to produce P and Q, but we also need a further item to be assembled to finish P.

Let's find out if this plant is profitable.

1. Product P
   Sales price = 90$
   Cost of raw material = 20$ + 20$ + 5$ = 45$
   Margin = 90$ – 45$ = 45$
   Market demand is 100 units, and the total profit margin of "P" is, therefore, 4500$.

2. Product Q
   Sales price = 100$
   Cost of raw material = 20$ + 20$ = 40$
   Margin = 100$ – 40$ = 60$

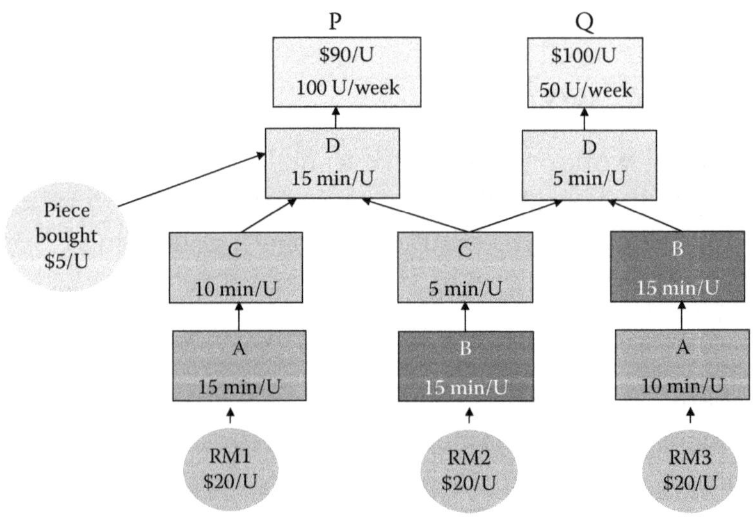

Figure 12.45 Production flow in the P & Q exercise.

Market demand is 50 units, and the total profit margin of Q is, therefore, 3000$.

Would we buy this company knowing that in one week the company costs equal 6000$ (net profit calculation)?

Well, the total profit margin is:

4500$ + 3000$ = 7500$.

If the weekly costs are 6000$, then there will be a profit of 1500$!

Unfortunately, it is not that simple.

We forgot one important thing: we cannot produce enough to satisfy market demand!

If we calculate the amount of total time used by each machine to produce 100 of P and 50 of Q, we will discover that the time used is as follows:

Machine A

15 × 100 min = 1500 min to do 100 P

then

10 × 50 min = 500 min to do 50 Q

For a total of 2000 min out of a maximum of 2400 min available (5 days of 8 h each)

Machine C and Machine D

15 × 100 min = 1500 min to do 100 P

then

5 × 50 min = 250 min to do 50 Q

For a total of 1750 min out of a maximum of 2400 min available

Machine B

15 × 100 min = 1500 min to do 100 P

then

30 × 50 min = 1500 min to do 50 Q

For a total of 3000 min out of a maximum of 2400 min available… IMPOSSIBLE!

So we have realized that we cannot meet market demand.

Let's identify the best mix to produce, then, and after that we can make a decision.

Which is the most profitable product?

The profit margin of P is 45$ and of Q is 60$. Clearly Q is the most profitable.

And what about production time? If we check the machine times we will see that product P takes a total time of 60 min to be produced whereas Q takes only 50 min.

Finally, if we consider how much raw material we need to keep in stock per product unit, we realize that the value of this is 45$ for one unit of P and only 40$ for one unit of Q.

We can conclude, on the basis of all three traditional accounting criteria, that the product we should concentrate on is Q!

So we do 50 of Q with a profit margin of 60$ each for a total of 3000$.

This production will take up 30 × 50 min of constraint time (machine B) leaving

200 – 1500 = 900 min available to do P.

As product P requires 15 min of constraint time we can make 900/15 = 60 units, which with a profit margin of 45$ each will give us 60 × 45$ = 2700$ extra.

Our total gain is then 3000$ + 2700$ = 5700$.

Taking into account that the weekly costs are 6000$, this leaves us with a not very satisfactory amount of –300$!

This is still not correct.

We have to forget about traditional cost accounting, and we have to use our common sense and some intelligence that we have acquired form the theory.

Using the traditional indicators of cost accounting, we have concentrated production on the product that takes up the most time from our scarce resource and therefore we have *artificially limited the maximum result we can achieve.*

As it is the constraint that determines the result, we have to focus on making sure that we make choices that maximize our earnings *on the constraint.*

If we calculate the profit margin earned per minute of constraint time for each product we can see that

with P we achieve $45/15 = $3 per minute of constraint
with Q we achieve $60/30 = $2 per minute of constraint

This tells us that we must maximize production of P!

We must therefore produce 100 P which will generate $45 × 100 = $4500

This requires 1500 min of time from machine B, leaving 900 min available to produce Q.

In 900 min we can do (on machine B) 900/30 = 30 Q

which will generate $60 × 30 = $1800

Our total earnings at this point are \$4500+ \$1800 = \$6300. If we subtract from this our weekly expenses, we are left with a net profit of \$300.

In the end, this company can be profitable!

12.7 Interacting Systemically along the Chain with Customers and Suppliers

Organizations, as we have said, are sets of interdependent components that work together to pursue a common goal. As we are adopting a systemic approach, we know that we *cannot* consider customers and suppliers as entities that are external to our system. So far we have focused our efforts on reducing variation *inside* the system, whether it is due to machines, processes, or interpersonal relations. What about our relationship with customers and suppliers?

If the entropy of a closed system always increases and we manage to decrease variation (entropy) inside the organization, entropy must increase somewhere else outside of it. This is what the second law of thermodynamics tells us. This means that, if we do not take into account customers and suppliers, we will inevitably run the risk of increasing entropy (variation) in our relationship with them. What can we do about it? There is a very specific approach from the TOC called "external constraint" that tackles the problem of the relationship with customers and suppliers in a very detailed way. It creates involvement among all the partners in the chain. The name "external constraint" refers particularly to customers and has to do with selling the "excess capacity" of an organization. However, the approach is equally useful with suppliers.

Let's see how it works.

12.7.1 Finding Agreements That Maximize Value for the Whole Chain

The question becomes, how can we find agreements that maximize profit for *all* the elements in the value chain, including customers and suppliers? What part do we play in that?

We already have powerful tools to do this with. These are the TP tools that we looked at in Chapter 10. Specifically, in our relationship with customers, when we are not selling everything that we could to the market, there is a completely generic and universal way to address the problem of selling all of our excess capacity.

We start by identifying the UDEs connected with

- Our market
- Our product or service
- Our organization in relation to sales

We can then build up the core conflict of external constraint. This core conflict summarizes the UDEs into a position:

D: Base the success of sales on negotiations essentially linked to price/volume and initiatives that are not strategically planned.

B: Not miss out on any opportunity.

The conflicting position is:

D′: Build a relationship with the client that is structured and mutually satisfying.

What is the need that the position D′ is trying to protect?

C: Have all the potential value of our work recognized.

The goal that is common to these two needs is:

A: Guarantee over time the economic sustainability of the organization by selling all the capacity available.

Let's look at these positions laid out in the conflict cloud format and interconnected through the various assumptions that underlie them (Figures 12.46 and 12.47).

When we invalidate the assumptions between D and D′ with a set of injections that respect the needs, we identify a complete, systemic pattern of behavior to overcome the problem of external constraint (Figures 12.48 and 12.49).

For each injection, there is a PRT to achieve the injection, and a TRT that details the tasks we have to carry out to achieve each IO of the PRT. This generic solution can be adopted and adapted by pretty much any organization.

12.7.2 Marketing and Sales as Systemic Activities (External Constraint)

The same steps that we saw from synchronous management apply to systemic marketing and sales. We identify the constraint (our ability to sell to the market); we exploit the constraint by making it work at 100% of

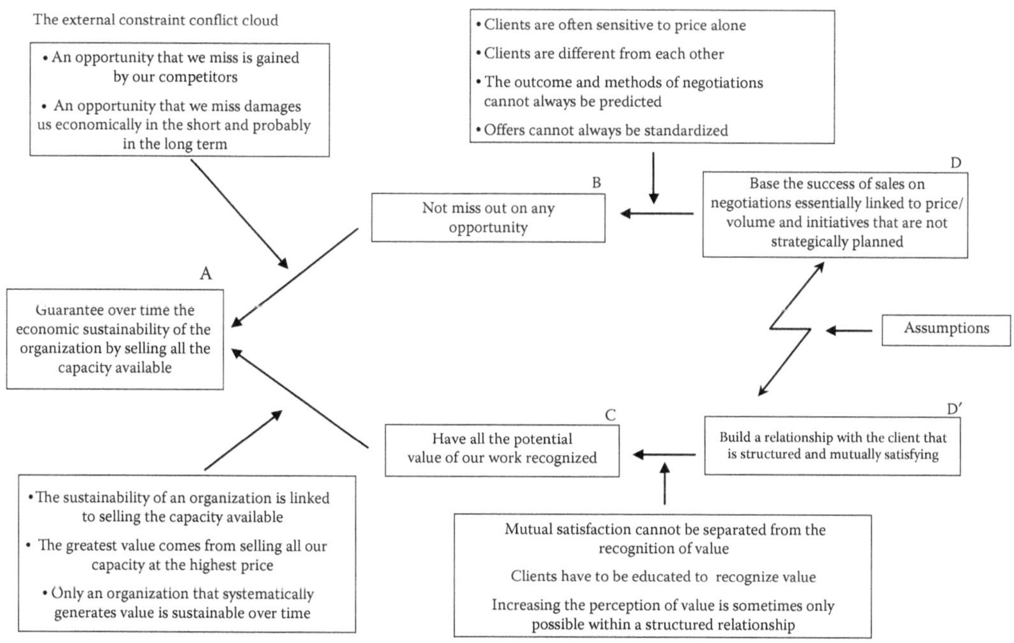

Figure 12.46 External constraint core conflict cloud.

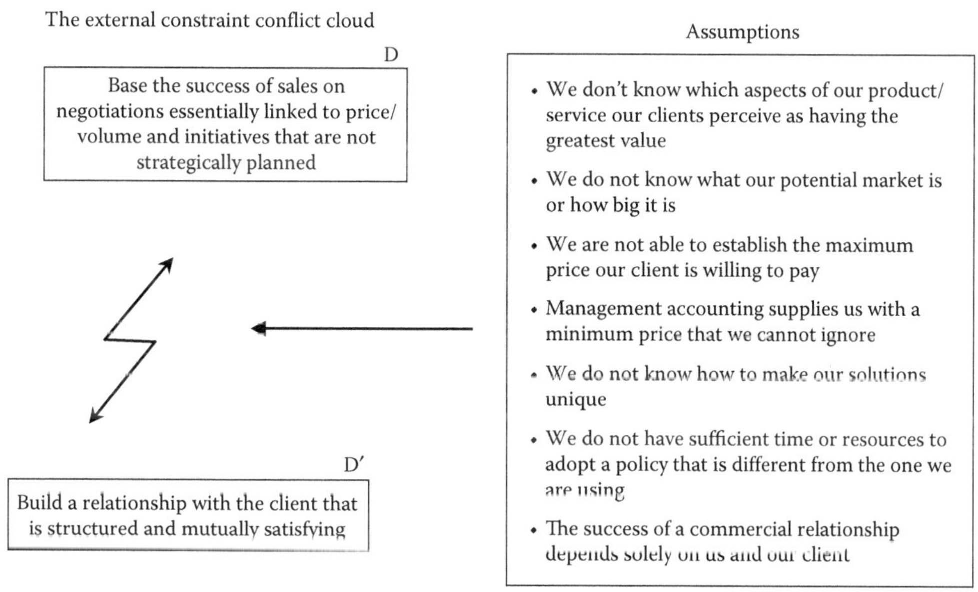

Figure 12.47 Assumptions between D and D′ for the external constraint core conflict.

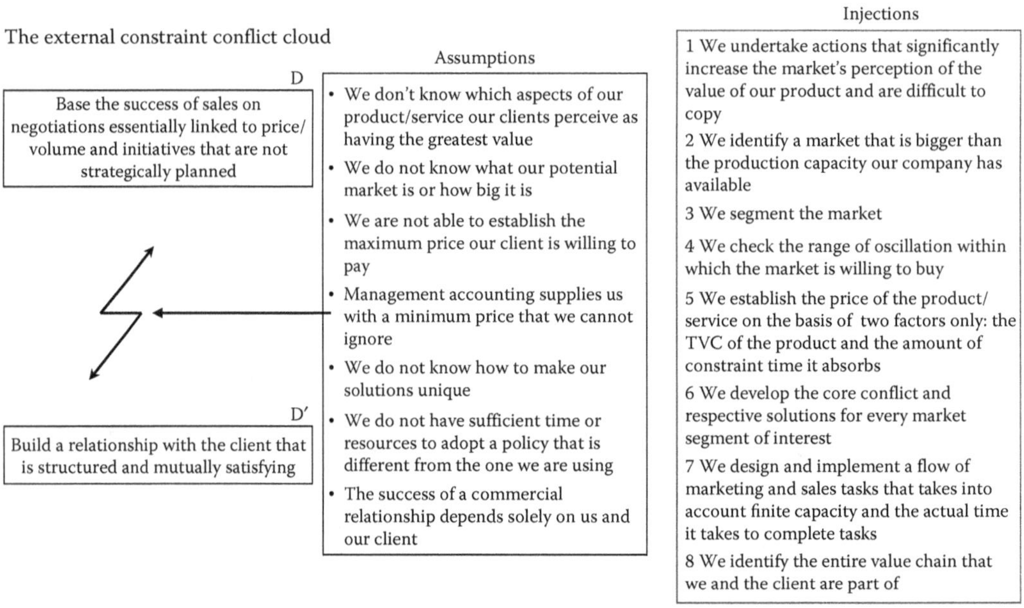

Figure 12.48 Injections to the external constraint conflict.

its capacity on the right client "mix"; we subordinate to the constraint by designing the process so that it supports the constraint; we "protect" the process by doing control charts of the cash generated by sales to understand the variation.

12.7.3 Win–Win Sales

It took us considerable time and effort to extract and rebuild all this detailed knowledge about external constraint from Goldratt's book "It's Not Luck," and we can say that after many years of exposure to it in practice, the external constraint algorithm works incredibly effectively. It does, however, require a considerable shift in mindset on the part of the sales team. The job is no longer about "selling at" people based on pure persuasion techniques, gift of the gab, and a price list. It becomes sales as part of detailed conversation with the customer about how everybody involved can make things better in a win–win way. It requires meticulous work in order to do the following:

- Identify the market through accurate segmentation
- Establish a price based on TVC (totally variable cost) involved in delivering the product/service and the constraint time it absorbs

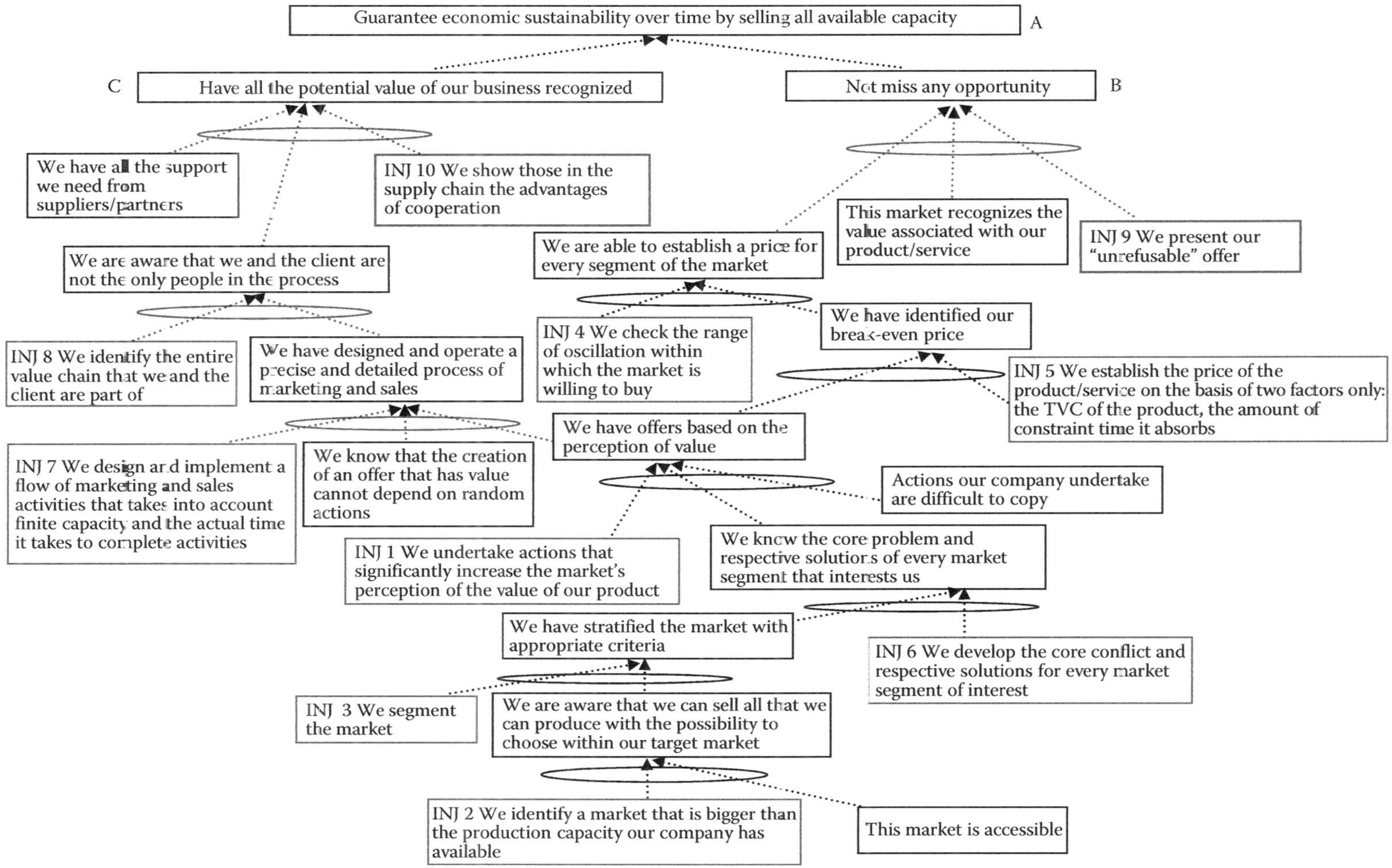

Figure 12.49 Injections to the external constraint conflict displayed within a future reality tree.

- Increase market awareness of the value of our offer
- Create an unrefusable offer for the client based on *their* UDEs and core conflict
- Show the entire chain involved the advantages of cooperation

When we present an offer, no matter how good it is, we must also be aware that customers and suppliers will go through the six levels of resistance that we saw in Chapter 10. Once again, by using the appropriate TP tool for each level or resistance, we can work with them in the direction of finding an agreement.

We need to SEE our customers and suppliers, the way we see a system. How can we do that? By understanding their point of view, and that means understanding their challenges (UDEs), and their legitimate needs regarding the market. Everybody has their own core conflict cloud and the more we know about it the more we understand about how we can all interact for mutual benefit.

Moreover, every time we present the customer with a new solution to an existing problem they are experiencing, what we actually do is create innovation. The essence of innovation is to remove a limitation, thus breaking through to a new and better reality and that is what the external constraint algorithm helps us to model and reproduce whenever we need to.

12.8 Continuous Breakthrough, Innovation, and Personal Development

It should be clear by this point that what we are advocating for leaders and managers is a continuous pattern of breakthrough and innovation, not just in organizational design but in everyday behavior. This level of focus and flexibility is required in today's complex reality.

Shifting away from traditional functions and hierarchy toward a systemic organization is a challenge in every sense: organizationally, psychologically, and emotionally. On the upside, and it's a major upside, it harbors the potential for a much more fulfilling relationship between people and their work, one of satisfying involvement. A leader in the twenty-first century will need to understand that the socioeconomics based on networks changes the behavioral and cognitive landscape and alters existing relationships among organizations and the inner needs of the people that are part of them.

A leader will have to have a good understanding of the neuroscience that informs how we think; business school will have to cater for that by linking the applied science that we have presented with its implications for our emotions and help students to tap into a higher level of self-consciousness. Such an elevated form of understanding of economics in the new realm of networks will lead to the fundamental questions of "ethics and wealth creation," a largely overlooked domain of knowledge, especially in business.

12.8.1 Creating the Right Network of Conversations through the Thinking Process Tools

Synchronization and variation are not only relevant to the mechanical process of an organization. They are equally relevant for human interaction. Indeed, we might even say that the lifeblood, or the links that connect the elements of an organization are the conversations for action that take place. The effective synchronization of these conversations is possible only if the basic processes that these conversations represent are statistically predictable in their outcomes. The question then becomes "How can we make the outcome of conversations predictable, how can we frame a human interaction based on language (as opposed to a one-way command to a machine) in a way that leads to predictable and actionable outcomes?"

We can replace the outdated hierarchical/functional structure with a much more organic design, based on managing variation and constraints, that reflects the intrinsically project-like nature of the work of organizations. This change is not cosmetic, it is transformational and it is rooted in a paradigm of cooperation, togetherness, and win–win. The new covenant that everyone in the organization (as well as the value chain in which the organization is embedded) must embrace requires a much higher ability to think, communicate, and act; it requires a new "wiring" in the way we measure, manage, and sustainably improve our efforts toward our goals.

The real challenge in bringing about a transformation based on a network of projects lies in the emotional and cognitive shift that needs to occur in the way people learn and use their knowledge as well as how they see themselves develop and interact in the workplace. The TP tools from TOC that we amply presented in Chapter 10 were developed by Dr. Goldratt to aid that shift. As we saw, Goldratt identified three major phases of change: (1) What to change; (2) What to change to; (3) How to make the change happen. He developed logical tools to support and facilitate each of the three phases. He did so as he realized how important it was to provide a strong cognitive support to combat the difficulties associated with change that people inevitably experience.

Far from being a mere technique, we have found these tools to be a critical element in transforming organizations into "thinking systems" because they foster in people the ability to see interdependencies, resolve conflicts, be empowered, work cooperatively, communicate effectively and, at a higher level, perceive (and act coherently upon) the oneness of the organization with its business and physical environment. Moreover, these tools can and should be used in all parts of the network, not just by leaders and managers. (Indeed, they are taught around the world to children in schools.)

To replace successfully a traditional hierarchy with a network of projects, we need to start from the very foundation of how we think, speak, and act. What we need to learn is how to seamlessly connect our uniquely human abilities to (a) systematically develop solutions to seemingly insurmountable conflicts (intuition), (b) understand the full spectrum of these solutions (analysis/understanding), (c) develop and implement a cohesive and coherent action plan to implement the solutions (knowledge), and (d) sustain the development of the "intelligent emotions" needed to harness complexity. The TP tools, when used regularly and consistently, allow people to develop the thinking, the systemic awareness, and the continuous breakthroughs they need to govern and thrive within complexity.

12.8.2 Intelligent Emotions

Using precise verbalization and simple diagrams, the TP tools help people to visualize the complex, highly nonlinear network of cause–effect relationships that make up reality, as we perceive it. They allow us to map the "conversations" that make up our cognitive horizon, and these conversations are in themselves a kind of network. Without appropriate tools, we have no way of grasping the reality of this network, how it defines the semantic boundary of our universe, and how it shapes our actions. Learning to use these tools takes some practice, but within days they allow people to focus and accelerate their work together. Over time, they enhance and fortify the faculties of the intellect that are responsible for the conception of new ideas (intuition), their full development through analysis (understanding) and the operational deployment of the actions needed to carry out the implementation of the fully analyzed idea (knowledge).

By linking these faculties, the TP tools enable a higher level of control over the interdependencies among these faculties, and this can often lead to conceiving breakthroughs that did not previously seem possible. The tools also have an important role in reducing variation in our thought processes by focusing our mental efforts toward a goal. They can greatly reduce variation in the way people communicate in an organization by providing a common language. Moreover, they help to harness the powerful forces represented by the emotions involved in the change process, and reinforce and engender the empathy required for collaborative work. The learning required through a change process can be very destabilizing on an emotional level because it continuously pushes forward the limits of our cognition. This means creating a gap between what we know and what we feel we can do, and this gap can be emotionally uncomfortable. In order to leverage the tension resulting from this gap in a positive way, we need to understand our emotions better and refine them. In this way, we can transform the potentially destructive power of emotions into a positive force for change. The TP tools help people to manage the interdependency of intellect and emotion in the change process. In this way, change can become the transformational effort that is required for any sustainable, as opposed to temporary, shift.

12.8.3 Involving the Individual at Every Level in the Organization

The starting point of any scientific investigation of the natural world is the understanding of the forces that act on and shape the subject under scrutiny. In a similar manner, when we seek to activate a transformational change in the way humans translate what they know into coherent behaviors, we have to understand the fundamental forces that act upon them and make them who they are. This is because these forces do influence the cognition process and limit or enhance the ability to learn.

The forces that shape how we learn and translate what we learn into consistent actions are not of the same kind that we study in physics. These forces shape individuals as well as organizations and they are responsible for the mental models that limit the way we perceive the world around us.

We can understand these forces only if we pierce the outer layer of the self and we connect with the energy within. This energy, we can call it "life," is what originates all of our faculties. Whether we look at this from a business, scientific, philosophical, or religious perspective, the issue of our inner energy is central for any development in our understanding of the human condition.

As we have seen with the two needs in the conflict cloud, there are forces that trigger all those actions aimed at *restraining* our behaviors and act on us in a way that makes us develop the need for the control of the environment we are in. The development of these forces is very heavily influenced by the way we experience the world, by the defining moments of our life and by the social and cultural fabric in which we develop our relationships. The forces that shape the fundamental need for control that is common to any person need to be harnessed and refined if we want to avoid being dominated by them. We see these forces in the B of the core conflict cloud.

On the other hand, there are forces that drive us to desire to *go beyond* our current state and see ourselves projected into a different and greater dimension of existence. This is the reflection of an often untapped level of consciousness and it is represented in the C of a core conflict cloud.

An expression of these forces can be found in two, seemingly conflicting desires that humans have, the harmonizing of which is central for a thriving organization. These are the desires to

- Belong and to bond (be part of a community)
- Develop our unique, distinct identity (be an individual)

How can the interests of the community (organization) and the individual (employee) all be fulfilled and served?

As Dr. Deming repeatedly pointed out, an individual's performance within an organization is the result of what the individual is capable of PLUS all the interaction and interdependencies in which they are immersed. As a result of these interactions, Dr. Deming rightly claims that assessing individual performances in an organization is both dangerous and useless.

An organization is different from a group of people waiting for a bus not because they do not have a common goal, they do, but because an organization ties the individuals and their actions within a network of interdependencies. These ties have the power to multiply the outcome of these actions or, sadly, debase it. These ties, in creating a unique way in which individuals interact, give origin to an entity, "the organization," that is subject, just like any individual, to the forces we described above. The modern lingo to describe the shaping impact of these forces is "corporate culture" and the job of HR seems to have become the development of appropriate metrics. Such metrics are almost invariably psychostatistical approximations that more often than not stifle innovation and beef up controlling syndromes.

From a systemic point of view, individual and organization are two aspects of the same thing and their interaction creates the emergent property of the organization as it behaves in the world. It is important to remember that

- Both community and the need for individuality are needed and necessary.
- A "unity" of the two is possible: a community needs to be made up of individuals that contribute their unique capabilities toward the goal.
- The organization and the individual must be nurtured and are of value in their own right.
- Each consists of the other: organizations can work in a way that fosters individual talent and contributions.
- Imperfect individuals are the elements of a continuously-improving-to-be-perfect community.

The most perfect community is made up of individuals exercising their individuality, developing their unique potential, and at the same time, the only way an individual can truly actualize their potential is through belonging to a healthy community. This is a place where meaningful work can be done.

The forces we have described above act upon the organization, guide its choices and reinforce the corporate culture; the individuals that are part of it become one with this culture, and the feedback loop "organization–individual" is constantly kept in motion. The knowledge and tools to create organizations that can foster quality, involvement, and flow in line with our legitimate ambition for meaningful work already exist and we hope to have illustrated a pathway toward an orderly acquisition of that knowledge (Figure 12.50).

12.8.4 Overview of the Essential Knowledge Pattern for Leaders and Managers

What follows is an overview of how the knowledge pattern we have described in the previous sections could be introduced and gradually absorbed through four phases. *All of the learning included adds direct value to any organization, without the need for writing any papers.* The TP tools from TOC are distributed throughout the learning phases.

Phase One: Building an awareness of systems:

1. An introduction to complexity
2. The systems view of an organization

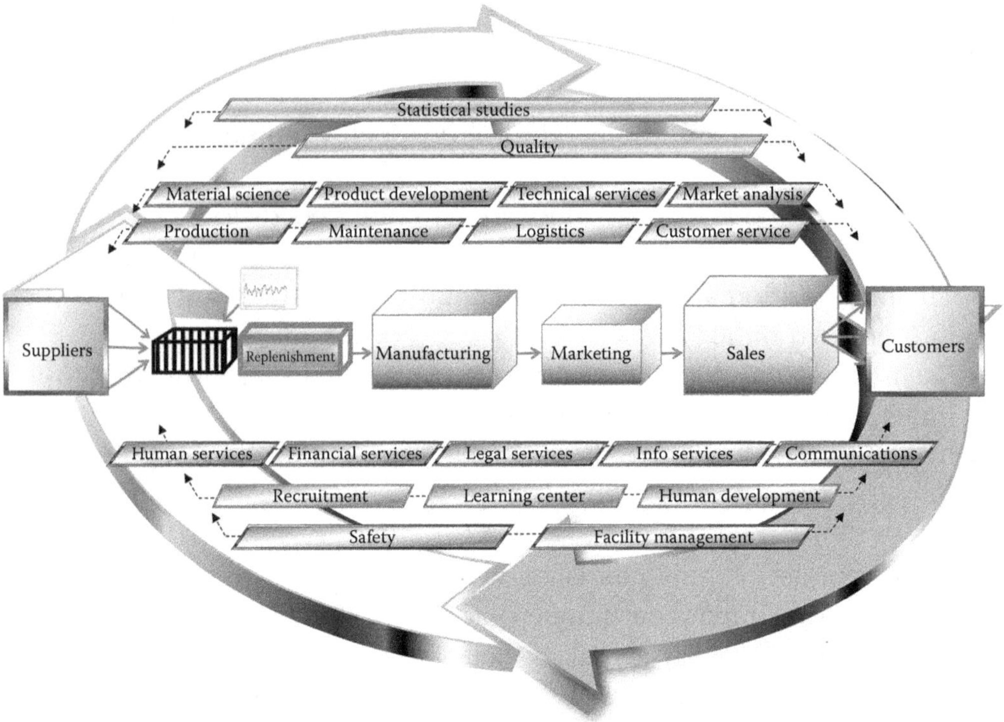

Figure 12.50 Model of an organization as a system built around the constraint of the replenishment phase.

3. An introduction to systemic thinking processes
   Conflict cloud
   Solving day-to-day conflicts
   Solving dilemmas
4. Processes and learning to map the system
5. Describing processes with a TRT

Phase Two: Managing the system to build quality, involvement, and flow:

1. Understanding variation
2. Measuring and monitoring variation through statistical process control
3. Managing an organization through a constraint (the simulator)
4. Buffer management
5. Systemic project management: Critical Chain
6. Planning projects using the PRT and TRT
7. Systemic management thinking skills:
   Core conflict cloud: from undesirable effects to the core conflict
   resolving the core conflict with systemic solutions (injections)

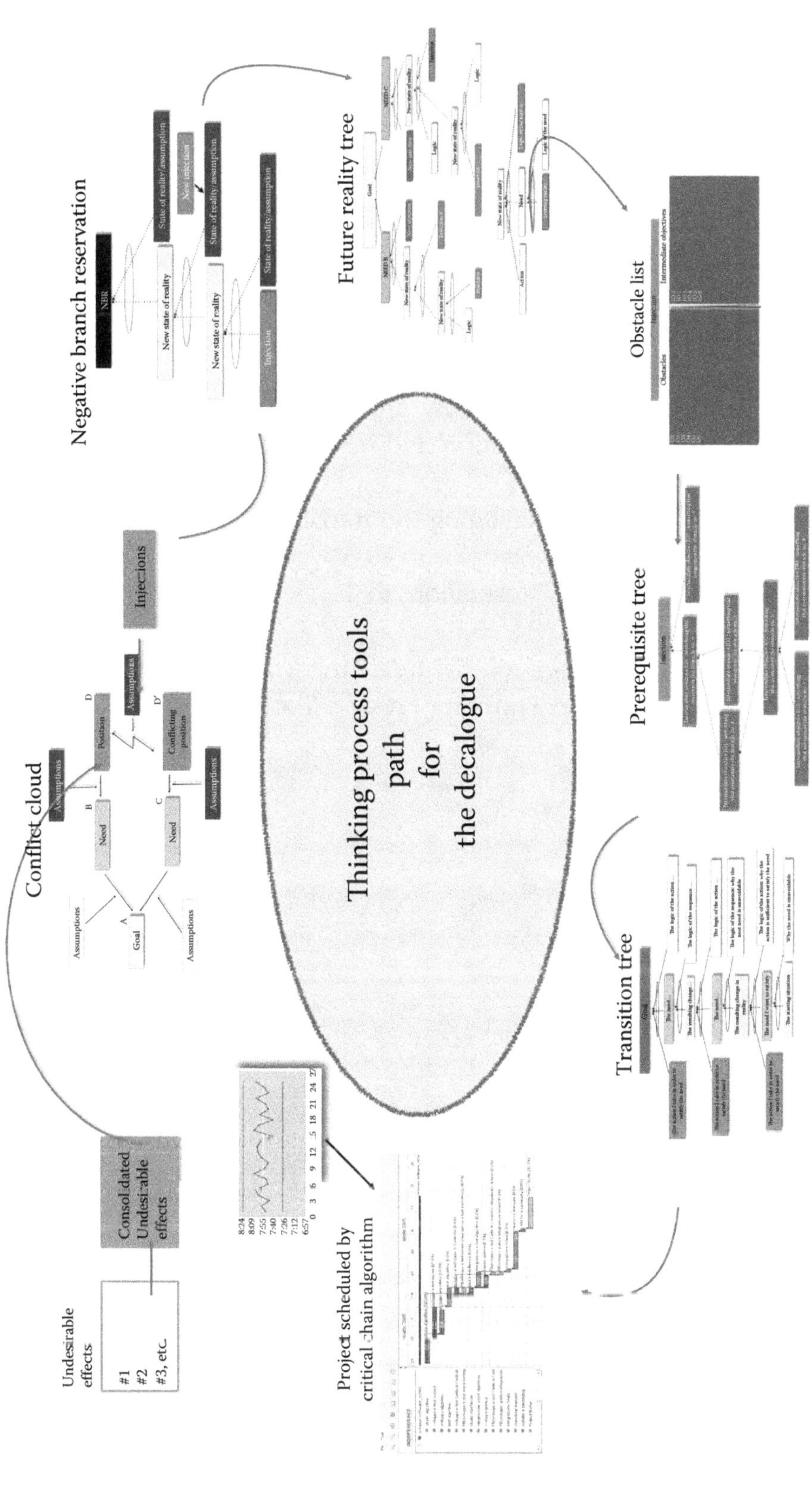

Figure 12.51 Complete cycle of application of the thinking process tools for a systemic organization.

Phase Three: Measures and measurement system for creating value through the value chain:

1. Understanding value and profitability: Product P & Q
2. Accounting for growth: throughput accounting (as previously stated, throughput accounting is not a replacement for traditional accounting but needs to be understood and applied in parallel in order to inform strategy and decisions)
3. Creating value through marketing and sales by selling the capacity of the system: external constraint
4. Continuous breakthrough and new product innovation: using the core conflict cloud to build innovation and new products

Phase Four: Planning and managing the transformation:

1. Building the path for transformation: FRT
2. Avoiding pitfalls: NBR
3. Highlighting the project-based nature of the Transformation through the application of all the Thinking Process Tools to the realization of a project (Figure 12.51).

Afterword: Economics, Innovation, and the Web

The last century has seen, along with unspeakable tragedies, the development of an unprecedented freedom to innovate, the tumbling down of physical and spiritual walls, and the birth of a new era of knowledge and information flow.

Dr. Deming advocated a *New Economics* to replace the "old" economics that has failed us. Deming's new economics is based on strong industry fuelled by an ever-growing speed of innovation and supported by a system of physical and cultural infrastructures provided by governments cognizant that their role is to facilitate the growth of talent, protect everybody's freedom to participate and make ventures, and help the vulnerable (children, elderly, sick, underemployed, and those living in poverty) to live with dignity. He was decades ahead of his time.

Today, post 9/11 and post global crisis that began in 2007, we live in an economic reality that is almost unrecognizable. No longer do we have economic cycles from boom to bust. We are immersed in new patterns of "long tails" for which nobody seems to have developed appropriate models, and yet economists continue to make predictions and offer recipes such as "austerity" that lack any basis in contemporary science. The new patterns that are emerging are precisely that: emergent properties of the unprecedented levels of interdependence that influence the outcome of all our interactions: social, political, and economic. Indeed, recent financial disasters and increasing problems linked to climate change have been triggered by the inability of leaders and governments to interpret data and behaviors systemically and therefore to adopt appropriate systemic solutions.

The more we understand the ever-increasing magnitude of the interdependencies we live in, the more we realize how inadequate current economic models are to inspire and guide our economic development. Ours is already (and much more in the near future) a completely interconnected economy, a network of interests and exchanges that is, albeit only partially, understandable only in light of the laws that govern networks. Leaders should be versed not just in these laws but more fundamentally in the organizational, philosophical, and cognitive paradigm that this economy shaped by networks entails.

Our chances of surviving and thriving are strictly related to our ability to understand complexity and to adopt a systems view, method, and tools to navigate within it. Our ability to innovate is the key to keeping up in the current market conditions. Innovation and technology can play an important economic role in the transformation away from the artificial strictures of a silo mentality if they focus on reducing current limitations. In other words, any innovation, especially technological, has its greatest value in diminishing some existing limitation in the way individuals and organizations lead their personal and socioeconomic life. Innovation, then, is the overcoming of a limitation, something that brings an unprecedented level of freedom to our lives because it makes that previous limitation disappear. Methodologically, we have a precise pattern to arrive at and continuously produce innovations. By using the conflict cloud meticulously, every time we invalidate a set of assumptions between the positions of D and D′ in the cloud with a solid injection, we produce a real and actionable innovation.

What we are saying, then, is that we believe that the video game *Angry Birds* is not an innovation, regardless of the money it generates and yes, we believe that *innovation* in marketing should cease to be the art and science of persuading people they need something they often do not have the money to buy.

The Web, arguably the most transformational technology of the last few decades, represents an innovation whose potential is largely untapped. Socially and economically we are moving rapidly towards a "network of networks" and such Web-mediated interactions must be shaped by trust and transparency. A whole new economy will emerge from secure and "real life"-inspired interactions on the Web. This "new" economy, fuelled by technologies that leverage the power of transparency and trustworthiness, will increase the sensitivity of everyone to issues like sustainability, ecology, inequality, and everything that pertains to the well functioning of this universal network of which we are all part. The new economy, in all its forms, requires leadership and management. All the foundational work, needed for management to become a new kind of science and, as such, a major contributor not just to wealth creation but also to wellbeing, already exists. Now is the time to apply that knowledge.

Bibliography

The following bibliography does not represent an exhaustive list of the resources that have been studied as part of the development of the work described in our book. This list does, however, provide a solid basis for an approach to all the aspects of systemic management.

Barabàsi, A. 2002. *Linked: The New Science of Networks*. Cambridge, Massachusetts: Perseus Publishing.

Barabàsi, A. and Reka, A. 1999. Emergence of scaling in random networks. *Science* 286(5439): 509–12.

Capra, F. 1996. *The Web of Life: A New Scientific Understanding of Living Systems*. New York: Anchor Books.

Capra, F. and Luisi, P. L. 2014. *The Systems View of Life: A Unifying Vision*. Cambridge: Cambridge University Press.

Cialdini, R. 2006. *Influence*. New York: Harper Business.

Corbett, T. 1998. *Throughput Accounting*. Great Barrington, Massachusetts: North River Press.

Deming, W. E. 1986. *Out of the Crisis*. Cambridge, Massachusetts: Massachusetts Institute of Technology Center for Advanced Engineering Study.

Deming, W. E. 1993. *The New Economics for Industry, Government, Education*. Cambridge, Massachusetts: Massachusetts Institute of Technology Center for Advanced Engineering Study.

Goldratt, E. 1998. *Essays on the Theory of Constraints*. Great Barrington, Massachusetts: North River Press.

Goldratt, E. M. 1984. *The Goal: A Process of Ongoing Improvement*. Great Barrington, Massachusetts: North River Press.

Goldratt, E. M. 1990. *The Haystack Syndrome: Sifting Information from the Data Ocean*. Great Barrington, Massachusetts: North River Press.

Goldratt, E. M. 1990. *What Is This Thing Called the Theory of Constraints and How Should It Be Implemented?* Great Barrington, Massachusetts: North River Press.

Goldratt, E. M. 1994. *It's Not Luck*. Great Barrington, Massachusetts: North River Press.

Goldratt, E. M. 1997. *Critical Chain*. Great Barrington, Massachusetts: North River Press.

Kahneman, D. and Tverski, A. 2013. *Thinking Fast and Slow*. New York: Farrar, Straus and Giroux.

Kahneman, D., Slovic, P. and Tverski, A. eds. 1982. *Judgement under Uncertainty: Heuristics and Biases*. Cambridge: Cambridge University Press.

Killian, C. S. 1988. *The World of W. Edwards Deming*. Washington, DC: CEE Press.

Lepore, D. 2011. *Sechel: Logic, Language and Tools to Manage any Organization as a Network*. Toronto: Intelligent Management Inc.

Lepore, D. and Cohen O. 1999. *Deming and Goldratt: The Decalogue*. Great Barrington, Massachusetts: North River Press.

Lepore, D., Montgomery, A. and Siepe, G. 2015. Managing complexity in organizations through a systemic network of projects. In: *Applications of Systems Thinking and Soft Operations Research in Managing Complexity*. Anthony Masys (ed.), pp. 35–69. Switzerland: Springer International Publishing.

Maci, G., Lepore, D., Pagano, S. and Siepe, G. 2008. Systemic approach to management: A case study. Poster presented at *5th European Conference on Complex Systems*, Hebrew University, Givat Ram Campus, Jerusalem, Israel, September 14–19, 2008.

Maci, G., Lepore, D., Pagano, S. and Siepe, G. 2008. Managing organizations as a system: The Novamerican case study. Poster presented at *International Workshop and Conference on Network Science*, Norwich Research Park, UK, June 23–27, 2008.

Mandelbrot, B. 1982. *The Fractal Geometry of Nature*. New York: W.H. Freeman.

Mandelbrot, B. and Hudson, R. L. 2004. *The Misbehavior of Markets: A Fractal View of Financial Turbulence*. New York: Basic Books.

Neave, H. 1990. *The Deming Dimension*. Knoxville, Tennessee: SPC Press. Deming of America. 1991. (Documentary) Cincinnati, Ohio: The Petty Consulting/ Productions

Shewhart, W. A. 1931. *Economic Control of Quality of Manufactured Product*. New York: van Nostrand Company Inc.

Shewhart, W. A. 1986. *Statistical Method from the Viewpoint of Quality Control*. Mineola, New York: Dover.

Watt, D. J. and Strogatz, S. 1998. Collective dynamics of "small world" networks. *Nature* 39: 440–42.

Wheeler, D. J. 1983. *Four Possibilities*. Knoxville, Tennessee: SPC Press; *Understanding Statistical Process Control*. 1992. Knoxville, Tennessee: SPC Press.

Wheeler, D. J. 1993. *Advanced Topics in Statistical Process Control*. 1995, Knoxville, Tennessee: SPC Press.

Wheeler, D. J. 1993. *Understanding Variation*. 1993. Knoxville, Tennessee: SPC Press.

Wheeler, D. J. 1998. *Avoiding Manmade Chaos*. Knoxville, Tennessee: SPC Press.

Wheeler, D. J. 1998. *Building Continual Improvement*. Knoxville, Tennessee: SPC Press.

Winograd, T. and Flores, F. 1987. *Understanding Computers and Cognition*. Boston: Addison-Wesley.

Index

F

G

H

I

K

L

P

Q

R

S

Authors

Dr. Domenico Lepore is an organizational scientist whose focus is on systemic management. He developed the Decalogue™ management methodology, together with Oded Cohen, world-renowned expert in the Theory of Constraints. The Decalogue provides a method for whole system transformation, from conventional, silo thinking to a systemic vision and practice.

Domenico and Oded coauthored the book *Deming-and-Goldratt: The Decalogue*, published in the United States in 1999 by Dr. Goldratt's publisher, North River Press. This was the first book on the Theory of Constraints to be published by North River Press that was not authored by Dr. Goldratt himself.

Domenico's background is in experimental physics and he has developed an original insight into the design of organizations as networks of projects. He has led successful implementations of the Decalogue in dozens of organizations for almost two decades and unique applications of Theory of Constraints to highly complex business environments.

Dr. Angela Montgomery has been partner of Domenico Lepore since cofounding their first Decalogue company in Milan, Italy in 1996. She has contributed over the years to the development, dissemination, and communication of the Decalogue methodology. Since moving to North America, she has developed a special interest in working with the Thinking Processes from the

Theory of Constraints to bring systemic thinking for positive change to organizations.

Angela has a PhD from London University in literature and science and has always been interested in crossing cultural boundaries. She blogs regularly for *Intelligent Management* and is author of the business novel *The Human Constraint* set in New York during the years of financial crisis. This book illustrates the power of a systemic approach and how it can point the way toward a sustainable future.

Dr. Giovanni Siepe has a background in theoretical physics. His research work on General Relativity has been cited internationally. He began working as a researcher in the microelectronics industry and he has developed patents in the field of power integrated circuits.

He continued his career as a manager in the plastics/material industry, where he developed significant experience in international marketing.

Over the last 12 years, Giovanni has worked at the advancement and dissemination of the Decalogue methodology, first leading the creation and development of algorithms and software (in liaison with the University of Salerno) for the management of complex systems, and later improving its integration with the use of Statistical Methods for managing organizational interdependencies.

For Product Safety Concerns and Information please contact our EU
representative GPSR@taylorandfrancis.com
Taylor & Francis Verlag GmbH, Kaufingerstraße 24, 80331 München, Germany

www.ingramcontent.com/pod-product-compliance
Lightning Source LLC
LaVergne TN
LVHW080650011026
849160LV00014B/1378
* 9 7 8 1 4 9 8 7 5 5 8 8 7 *